W0234872

TEXTS AND READINGS
IN MATHEMATICS

3

Representations of Finite Groups

Texts and Readings in Mathematics

Advisory Editor
C. S. Seshadri, Chennai Mathematical Institute, Chennai.

Managing Editor
Rajendra Bhatia, Indian Statistical Institute, New Delhi.

Editors
R. B. Bapat, Indian Statistical Institute, New Delhi.
V. S. Borkar, Tata Inst. of Fundamental Research, Mumbai.
Probal Chaudhuri, Indian Statistical Institute, Kolkata.
V. S. Sunder, Inst. of Mathematical Sciences, Chennai.
M. Vanninathan, TIFR Centre, Bangalore.

Representations of Finite Groups

C. Musili
University of Hyderabad
Hyderabad

 HINDUSTAN
BOOK AGENCY

Published by
Hindustan Book Agency (India)
P 19 Green Park Extension
New Delhi 110 016
India

email: info@hindbook.com
www.hindbook.com

Copyright © 1993, Hindustan Book Agency (India)

Digitally reprinted paper cover edition 2011

No part of the material protected by this copyright notice may be repro-
duced or utilized in any form or by any means, electronic or mechanical,
including photocopying, recording or by any information storage and
retrieval system, without written permission from the copyright owner, who
has also the sole right to grant licences for translation into other languages
and publication thereof.

All export rights for this edition vest exclusively with Hindustan Book
Agency (India). Unauthorized export is a violation of Copyright Law and is
subject to legal action.

ISBN 978-93-80250-18-2

To

My Teachers

C.S. SESHADRI and **M.S. NARASIMHAN**

on their

Sixtieth   Birthdays

Preface

This book[1] contains FOUR parts. The first two parts consist of a revised and expanded version of *Class Notes* of a *core* course of lectures I gave to the M.Phil.(Math) students of the University of Hyderabad, several times in the past decade, especially in the years 1987–88 and 1989–90. The remaining two parts are of an *elective* course of lectures given in 1987–88 which also formed the base for the expository dissertation for the M. Phil. degree of my student G. Srinagesh (cf. [48], cited in the bibliography).

The purpose of this lecture notes is *multifold*. The FIRST and foremost is to give an elementary introduction to the basic concepts of the theory of *ordinary* representations of finite groups (i.e., representations over algebraically closed fields of characteristic 0), with a minimum of prerequisites, making the theory accessible to students with just one semester exposure to the rudiments of Linear Algebra, Groups, Rings and Modules. This is done in the first two parts (i.e., Chapters 1 to 4, with the preliminaries being recalled in Chapter 1). Burnside's theorem on the solvability of groups of order $p^a q^b$ is included in Chapter 3, as a classical application of the theory to a concrete problem on groups.

The SECOND which is also the main theme of this exposition is to be able to do the theory rather explicitly for the important special case of the *symmetric* group S_n of permutations on n letters, with as little of the technical preparation as possible. The group S_n is so rich in structure that it allows the constructions to be done *as if by hand* and that too by more than one method. The several apparently different approaches, namely, **I.** FROBENIUS, **II.** FROBENIUS-YOUNG, **III.** SPECHT and other **IV.** ABSTRACT methods, all lead in the end to the same concrete realisations of the irreducible representations of S_n. Moreover, the interplay of ideas from one method to another makes the study rewarding and facilitates an appreciation of the theory even by a beginner. This is done in Chapter 5, the first three methods I, II and III above, being presented completely independent of Chapter 4.

---

[1]Written on LaTeX

The THIRD aspect is to use the preparatory material of the first two parts coupled with the S_n-theory to do the same for some other important special groups, namely, the Alternating group A_n (Chapter 6), the Hyperoctahedral groups B_n (Chapter 7) and D_n (Chapter 8). The case of B_n goes mostly (though not entirely) imitating the pattern for S_n. Once the job is done for A_n, almost the same is repeated for D_n since the pair (B_n, D_n) is quite similar to (S_n, A_n). While there is vast literature available on S_n and A_n, both in papers and books, I am not aware of any book containing the material on B_n and D_n.

The FOURTH aim is to prepare the reader for being in a position to look *around* and *ahead* so as to explore a lot more interesting frontiers that are beyond the scope of this book (and not even touched here). (Several possible further directions are kept as guides in compiling the not so extensive bibliography given at the end). A couple of them, for instance, are the following.
1. Representations of Weyl groups: What we have done here covers just the case of *Classical* Weyl groups, i.e., S_n (for the *Linear* groups GL_n or SL_n), B_n (for the *Symplectic* group Sp_{2n} or the odd *Orthogonal* group SO_{2n+1}) and D_n (for the even *Orthogonal* group SO_{2n}). What remains is the case of the five *Exceptional* types (E_6, E_7, E_8, F_4 and G_2). Good amount of material is available in [7], [33], [34], [47], etc.
2. Structre of the " Zero Weight" spaces of the irreducible representations of a "Linear Algebraic/Lie Group" G, as representations of the Weyl group $W(G)$ of G. It seems as though much remains to be explored in this direction. See [2], [8], [13], [17], [19], [20], [30], etc.

The main text does not depend on the exercises in a serious way, enabling self study. Nevertheless, they are class tested and are found to help a better understanding of the material. Ability to correctly sort out the True/False statements with proper justification is found to be a quick test of assimilation of the concepts. A glance at the table of contents gives a fairly good idea of the organisation of the material. Dependence of the chapters is progressively linear *except* that Chapters 5 and 7 do not depend on 4 and 6 respectively. The labelling for cross references is self explanatory. For instance, (3.10.7) refers to Chapter 3, section 10 and item 7. End of a proof is signalled by $\Diamond$ and of a section by ■

Generally speaking, it is observed that a student beginning to learn a new topic looks for a book that is *thin in volume but generous in details* of the basic material covered. The present volume is intended to meet this requirement, as vindicated by many users of these notes in their draft form. I would like to record my appreciation of the strong desire of some of them to make these available for a wider audience.

The draft version was revised and enlarged during May–June 1991 at the Tata Institute of Fundamental Research, Bombay, thanks to the kind hospitality of the School of Mathematics. Subsequently, based on the critical study and constructive suggestions offered by several of my colleagues at TIFR (Bombay), principally Dipendra Prasad, part of the material was reorganised and some of the proofs were simplified, resulting in the present form. I express my deep gratitude to them for their time and interest in this endeavour. Some of the final touches were carried on at the School of Mathematics of the SPIC Science Foundation, Madras, thanks to the invitation and opportunity provided to me during January–February 1992.

It gives me great pleasure to say that this work is meant as a small token of my indebtedness to my inspiring teachers who not only initiated me to many a topic decades ago but also enabled me to appreciate them even as a beginner.

Hyderabad,
1 November 1992. C. Musili

Table of Contents

Part I

THE STRUCTURE OF
SEMI-SIMPLE RINGS

Part II

REPRESENTATIONS OF
FINITE GROUPS

Part III

REPRESENTATIONS OF THE
SYMMETRIC AND ALTERNATING GROUPS

Chapter 5 : Representations of the Symmetric Group S_n

Chapter 6 : Representations of the Alternating Group A_n

Part IV

REPRESENTATIONS OF THE
HYPEROCTAHEDRAL GROUPS B_n AND D_n

Chapter 7 : Representations of the Hyperoctahedral Group B_n

Chapter 8 : Representations of the Hyperoctahedral Group D_n

Glossary of Notation

$\mathrm{Hom}_{R-\mathbf{alg}}(A,B)$	R–algebra homomorphisms from A to B	17
$\mathrm{Hom}_R(M,N)$	R–linear homomorphisms from M to N	5
$H^g = gHg^{-1}$	g–conjugate of H	124
$h(G)$	Number of conjugacy classes of G	71
h_λ	Hook–length of T_λ	177
$h_{(\lambda,\mu)}$	Hook–length of $T_{(\lambda,\mu)}$	217
$\mathrm{Irr}_K(G)$	$\begin{cases} \text{A complete set of mutually} \\ \text{ineqt. irred. repns. of G over } K \end{cases}$	95
I_W	Inertia group of a repn. W	125
ιB_n	Trivial character of B_n	201
ιS_n	Trivial character of S_n	147
$\mathrm{Ind}(\ \ )\!\uparrow_H^G$	Induced repn. from H to G	116
$\langle\ ,\ \rangle_G$	Inner product on the functions on G	89
$J(R)$	Jacobson radical of a ring R	14
perm_H	Permutation representation on G/H	84
reg.	Regular representation of G	84
$\ell_R(M)$	Length of an R–module M	11
$\lambda \vdash n$	$\lambda = (\lambda_1,\cdots,\lambda_r)$, a partition of n	147
$\lambda' \vdash n$	Partition conjugate to $\lambda \vdash n$	152
λ–tab.	Young tableaux corresp. to $\lambda \vdash n$ 152,	154
$[\lambda]$	Irred. repn. of S_n corresp. to $\lambda \vdash n$	181
$(\lambda,\mu) \models n$	Complementary partitions of n	202
(λ,μ)–diag.	Young diagram corresp. to $(\lambda,\mu) \models n$	205
(λ,μ)–tab.	Young tableaux corresp. to $(\lambda,\mu) \models n$	205
$[\lambda,\mu]$	Irred. repn. of B_n corresp. to $(\lambda,\mu) \models n$	219
$M_n(R)$	Ring of $n \times n$ matrices over R	4
$M \oplus N$	Direct sum of modules M and N	6
$M \otimes_R N$	Tensor product of M and N over R	20
$\mathbf{N},\ \mathbb{N}$	Positive integers	4
$N \bullet H$	Semi–direct product of H by N 115,	138
$N(R)$	Nil radical of a ring R	14
ηB_n	sgn^+ character of B_n	201

Part I

THE STRUCTURE OF
SEMI–SIMPLE RINGS

Chapter 1

Preliminaries

In what follows, some basic knowledge of Rings and Modules is assumed. By a *ring R*, we mean a ring R with unity, commutative or not. The *unity element* of R is denoted by 1. By an *ideal* of R, we mean a left/right/2–sided ideal, as the case may be (as specified in, or clear from the context). If two or more ideals are referred to at the same time, they are all assumed to be ideals of the same type. By an *R–module M*, we mean a *unitary left* (or right) R–module. We recall a few elementary properties of rings and modules, needed in the sequel, and set the notation. (Cf. [23], [31], [40], etc., for details.)

1.1 Rings and Modules

1.1.1 Opposite ring: Given a ring $(R, +, \cdot)$, let R^0 or R^{op} (read as R–opposite), be the same set R. With the same addition $(+)$ as in R, define multiplication $(*)$ on R^0 to be $a * b = b \cdot a$ for all $a, b \in R^0$. Under these operations, R^0 is a ring, called the *ring opposite* to R or the *opposite ring* of R. It is obvious that we have
(i) $(R^0)^0 = R$ and **(ii)** $R = R^0$ if and only if R is commutative.

1.1.2 Characteristic of a ring: Given a ring R, by the *characteristic* of R, denoted by Char R, we mean the least positive integer n such that $na = 0$ for all $a \in R$ if such an n exists; otherwise, it is defined to be 0. We note the following.
1. Char $R = 0$ if and only if the additive order of 1 is infinite, i.e.,

$n1 \neq 0$ for all $n \in \mathbb{N}$.

2. Char $R = n \neq 0$ if and only if the additive order of 1 is finite and is equal to n.

3. Let $P = \{n1 \mid n \in \mathbb{Z}\}$ be the subring of R generated by 1, called the *prime subring* of R. Then we have the following.

(i) Char $R = 0 \iff P$ is infinite and

(ii) Char $R = n \neq 0 \iff P$ is finite and has exactly n elements.

4. The characteristic of an *integral domain* (in particular, of a *division ring* or a *field*) is either 0 or a prime number.

5. Char $M_n(R) =$ Char R where $M_n(R)$ is the *ring of all $n \times n$ matrices with entries in R*.

6. Char $R =$ Char $R[X] =$ Char $R[X, X^{-1}]$ where $R[X]$ is the *ring of polynomials* and $R[X, X^{-1}]$ is the *ring of Laurent polynomials* in the variable X with coefficients in R.

7. Char $\mathbb{Z}_n = n$ where $\mathbb{Z}_n = \mathbb{Z}/n\mathbb{Z} = \{\overline{0}, \overline{1}, \cdots, \overline{n-1}\}$ is the *ring of integers modulo n* (for a positive integer n).

1.1.3 Zorn's Lemma: *A partially ordered non–empty set in which every chain (i.e., a totally ordered subset) is bounded above (resp. below) has a maximal (resp. minimal) element.*

Using this, one proves the existence of maximal ideals in a ring R. In fact, given an ideal $I \neq R$, there exists a maximal ideal $M \supseteq I$ of the same kind as I (left/right/2–sided). (This is not true if maximal is replaced by minimal or if R is without 1.)

1.1.4 Simple ring: A non–zero ring R is said to be a *simple ring* if R has no 2–sided ideals other than (0) and R.

Example: By Ex.(1.11.7) below, $M_n(D)$ $(n \geq 1)$ is a simple ring for any division ring D. (See (2.5.2) below, for the converse (under an additional hypothesis)).

1.1.5 Local ring: A ring R is called a *local ring* if the set of all non–units in R is an ideal. In fact, this ideal can be seen to be the unique maximal left/right/2–sided ideal of R and that the quotient ring is a division ring.

1.1.6 Remark: The characteristic of a local ring is either 0 or a power of a prime. In particular, $\mathbb{Z}/n\mathbb{Z}$ is local if and only if n is a power of a prime.

1.1.7 Semi–local ring: A ring having only finitely many maximal ideals is called a *semi-local* ring.

Notation: 1. Given rings R and S, the set of all *ring homomorphisms* from R to S is denoted by $\text{Hom}_{\text{rings}}(R, S)$.

2. Given R–modules M and N, the set of all *R–linear homomorphisms of modules* from M to N is denoted by $\text{Hom}_{R-\text{mod}}(M, N)$ or simply $\text{Hom}_R(M, N)$.

3. The set $\text{Hom}_R(M, M)$ of all *R–linear endomorphisms* is also denoted by $\text{End}_R(M)$.

Note: 1. $\text{Hom}_R(M, N)$ is an abelian group under pointwise addition of maps. In general, it is not an R–module. However, it is an R–module if R is commutative.

2. $\text{End}_R(M)$ is a ring with 1 under point–wise addition and composition of maps. It is a subring of $\text{End}_{\mathbb{Z}}(M)$.

3. Let V be a *left vector space* (i.e., a left module) over a division ring D. If V is finite dimensional, then the ring $\text{End}_D(V)$ of all D–linear endomorphisms of V is isomorphic to the ring $M_n(D)$ where $n = \dim_D(V)$. (Prove this.)

1.1.8 Finitely generated module: An R–module M is said to be *finitely generated* over R if there is a finite subset X of M such that M is the submodule *generated* (i.e., *spanned*) by X, i.e., if $X = \{x_1, \cdots, x_r\}$, then we have $M = \{\sum_{i=1}^{r} a_i x_i \mid a_i \in R\}$.

Using Zorn's lemma, one proves the existence of maximal submodules in a finitely generated non–zero module.

1.1.9 Free module: An R–module M is called a *free module* if M has a *basis* B, i.e., a linearly independent subset B of M such that M is spanned by B over R. This means that every element $x \in M$ can be written *uniquely* as $x = \sum_{b \in B} \lambda_b b$, with $\lambda_b \in R$ and $\lambda_b = 0$ except for finitely many b's (i.e., x is a finite linear combination of elements in B, the scalars being unique for x).

1.1.10 Simple module: A module M is called a *simple module* if
(i) $M \neq (0)$ and **(ii)** the only submodules of M are (0) and M.

1.1.11 Remark: A module M is simple $\iff$ M is generated by
any non-zero element of M $\iff$ $M \simeq R/I$ for some maximal left
ideal I of R.

1.1.12 Schur's Lemma: *Suppose N and M are two simple
R-modules. Then any R-linear map $f: M \to N$ is either 0 or an
isomorphism. In particular, $D = End_R(M)$ is a division ring.*

1.1.13 Sum of submodules: Given submodules P and Q of a
module M, the *sum of the submodules* is defined as the smallest submodule of M containing (or generated by) P and Q and it is simply
$P + Q = \{x + y \mid x \in P, y \in Q\}$.

This concept makes sense for any family (finite or not) $\{P_\alpha\}_{\alpha \in I}$ of
submodules of M, i.e.,

$$\sum_{\alpha \in I} P_\alpha = \{\sum_{\alpha \in I} x_\alpha \mid x_\alpha \in P_\alpha, x_\alpha = 0 \text{ except for finitely many } \alpha's\}.$$

This is the smallest submodule of M containing all the P_α's.

1.1.14 Direct sums: Given modules M and N, consider the
Cartesian product module $P = M \times N$ and observe that P con-
tains isomorphic copies of M and N as submodules, namely, $M = \{(x, 0) \in P \mid x \in M\} \subseteq P$ and $N = \{(0, y) \in P \mid y \in N\} \subseteq P$. The
sum of the submodules M and N in P is called the *external direct
sum of the modules* or simply the *direct sum of the modules* M and
N. This is denoted by $M \oplus N$.

This sum is direct in the sense that every element $z = (x, y) \in M \oplus N$
can be written as the sum $z = (x, 0) + (0, y) = x + y$ of *unique* elements
$x \in M$ and $y \in N$, or equivalently, $P = M + N$ with $M \cap N = (0)$.

A module P is called an *internal direct sum* or simply a *direct sum*
of a family of submodules $\{P_\alpha\}_{\alpha \in I}$ if $P = \sum_{\alpha \in I} P_\alpha$ and every element
$z \in P$ can be written *uniquely* as $z = \sum_{\alpha \in I} x_\alpha$, $x_\alpha \in P_\alpha$, $x_\alpha = 0$ except
for finitely many α's. We write $P = \oplus_{\alpha \in I} P_\alpha$.

We have $\oplus_{\alpha \in I} P_\alpha \subseteq \prod_{\alpha \in I} P_\alpha$ and *equality holds* $\iff$ *I is finite.*

1.1.15 Direct summand: A submodule P of a module M is said to be a *direct summand* or simply a *summand* of M if there exists a submodule Q of M such that $M = P \oplus Q$. Such a Q is called a *supplement* of P.

Note that a submodule need not be a summand or if it is, its supplement need not be unique. On the other hand, in a vector space, every subspace is a summand. This is a special case of a class of modules in each of which every submodule is a summand (§2.1 below). ∎

1.2 Artinian Modules

1.2.1 Theorem: *For an R-module M, the following are equivalent.*
1. *Descending chain condition (d.c.c) holds for M,*
i.e., any descending chain of submodules of M, say

$$M_1 \supseteq M_2 \supseteq \cdots \supseteq M_n \supseteq \cdots \supseteq \cdots$$

is stationary in the sense that $M_r = M_{r+1} = \cdots$ for some r. (We write this as $M_r = M_{r+1}, \forall\, r \gg 0$.)
2. *Minimum condition holds for M, in the sense that any non-empty family of submodules of M has a minimal element.*

1.2.2 Artinian module: A module M is called *Artinian* if d.c.c (or equivalently, the minimum condition) holds for M.

1.2.3 Remark: Minimal submodules exist in a non-zero Artinian module because a minimal submodule is simply a minimal element in the family of all non-zero submodules of M.

1.2.4 Examples: 1. A module which has only finitely many submodules is Artinian. In particular, finite abelian groups are Artinian.
2. Finite dimensional vector spaces are Artinian (for dimension reasons) whereas infinite dimensional ones are *not* Artinian.
3. Infinite cyclic groups are *not* Artinian. For instance, $\mathbb{Z}$ has a non-stationary descending chain of subgroups, namely,

$$\mathbb{Z} = (1) \supset (2) \supset (4) \supset \cdots \supset (2^n) \supset \cdots \supset \cdots$$

1.2.5   Theorem: 1. *Submodules and quotient modules of Artinian modules are Artinian.*
2. *If a module M is such that it has a submodule N with both N and M/N are Artinian, then M is Artinian.*

1.2.6 Corollary: *Every non–zero submodule of an Artinian module contains a minimal submodule.* (Obvious by (1.2.3).)

1.2.7 Corollary: *Sums and direct sums of finitely many Artinian modules are Artinian.*

1.2.8 Remark: Direct sum of an infinite family of non–zero Artinian modules is *not* Artinian.   ■

1.3   Noetherian Modules

A major part of this section (but not entirely) is dual to the previous one, by respectively changing "descending", "minimum", etc., to "ascending", "maximum", etc.

1.3.1 Theorem: *For an R–module M, the following are equivalent.*
1. *Ascending chain condition* (a.c.c) *holds for M, i.e., any ascending chain of submodules of M, say*

$$M_1 \subseteq M_2 \subseteq \cdots \subseteq M_n \subseteq \cdots \subseteq \cdots$$

is stationary *in the sense that* $M_r = M_{r+1} = \cdots$ *for some r.* (*We write this as* $M_r = M_{r+1}, \forall\, r \gg 0$.)
2. *Maximum condition holds for M in the sense that any non–empty family of submodules of M has a maximal element.*
3. *Finiteness condition holds for M in the sense that every submodule of M is finitely generated.*

1.3.2 Noetherian module: A module M is called *Noetherian* if a.c.c (or equivalently, the maximum condition or the finiteness condition) holds for M.

Note: **1.** While a Noetherian module is finitely generated, a finitely generated module need not be Noetherian.
2. The "finiteness condition" has no parallel in the Artinian case.

1.3.3 Remark: Maximal submodules exist in a non–zero Noetherian module because a maximal submodule is simply a maximal element in the family of all submodules of M *excluding* M.

Note: Maximal submodules exist in any finitely generated non–zero module (Noetherian or not). (This is a simple consequence of Zorn's lemma applied to the family of all submodules of M other than M.)

1.3.4 Examples: **1.** A module which has only finitely many submodules is Noetherian. In particular, finite abelian groups are Noetherian as modules over $\mathbb{Z}$.
2. Finite dimensional vector spaces are Noetherian (for dimension reasons) whereas infinite dimensional ones are *not* Noetherian.
3. Infinite cyclic groups are Noetherian because every subgroup of a cyclic group is again cyclic (but we saw in (1.2.4)(3) above, that they are not Artinian).

1.3.5 Theorem: **1.** *Submodules and quotient modules of Noetherian modules are Noetherian.*
2. *If a module M is such that it has a submodule N with both N and M/N are Noetherian, then M is Noetherian.*

1.3.6 Corollary: *Every non–zero submodule of a Noetherian module is contained in a maximal submodule.* (Obvious by (1.3.3).)

1.3.7 Corollary: *Sums and direct sums of* finitely many *Noetherian modules are Noetherian.*

1.3.8 Remark: Direct sum of an infinite family of non–zero Noetherian modules is *not* Noetherian.

1.3.9 Exercises: Construct examples to justify the following statements. (See also (1.8.5) below.)
1. *An Artinian module need not be finitely generated.*
2. *Maximal submodules need not exist in an Artinian module.*

3. *An Artinian module need not be Noetherian.*
4. *A finitely generated module need not be Noetherian.*
5. *Minimal submodules need not exist in a Noetherian module.*
6. *A Noetherian module need not be Artinian.*
7. *Some modules are neither Artinian nor Noetherian.* ■

1.4   Modules of Finite Length

1.4.1 Composition series: By a *composition series* of a non–zero module M, we mean a finite descending chain of submodules of M starting with M and ending with (0), say

$$M = M_0 \supset M_1 \supset \cdots \supset M_m = (0),$$

such that the successive quotients M_i/M_{i+1} are *simple* (1.1.10) for all $i, 0 \le i \le m - 1$. The integer m is called the *length* of the series. (It is one less than the number of terms in the series.)

Note: A composition series is also called a *Jordan–Hölder filtration* or simply a *filtration* and the *simple quotients* M_i/M_{i+1} are also called the *composition factors* of the filtration.

1.4.2 Remark: For a non–zero module M, a composition series may or may not exist. If one exists, we notice that M would have a simple (i.e., a minimal) submodule M_{m-1} and a simple quotient M_0/M_1 (equivalently, M_1 a maximal submodule).

1.4.3 Examples: **1.** A vector space V having a finite basis $\{v_1, \cdots, v_m\}$ has a composition series of length m, namely,

$$V = V_0 \supset V_1 \supset \cdots \supset V_m = (0),$$

where $V_i = $ span of $\{v_{i+1}, v_{i+2}, \cdots, v_m\}$ for all $i,\ 0 \le i \le m$ with $V_m = (0)$. However, a vector space having an infinite basis cannot have a composition series.
2. A finite abelian group has a composition series.
3. An infinite cyclic group cannot have a composition series since it has no minimal submodules.

1.4.4 Module of finite length: A module is called a *module of finite length* if it is either zero or has some composition series.

1.4.5 Theorem: *A module is of finite length* $\iff$ *it is both Artinian and Noetherian.*

1.4.6 Theorem: 1. *Submodules and quotient modules of a module of finite length are modules of finite length.*
2. *If a module M has a submodule N such that both N and M/N are of finite length, then M is of finite length.*

1.4.7 Theorem (Jordan–Hölder): *Any two composition series of a non–zero module are equivalent in the sense that both have the same length and the same simple quotients upto order and isomorphisms. To be more precise, let $M = M_0 \supset M_1 \supset \cdots \supset M_m = (0)$ and $M = N_0 \supset N_1 \supset \cdots \supset N_n = (0)$ be any two composition series for M. Then* **(i)** *$m = n$ and* **(ii)** *for each i, $0 \le i \le m - 1$, $\exists\, j = j(i)$, $0 \le j \le n - 1$ such that $M_i/M_{i+1} \simeq N_j/N_{j+1}$ and vice–versa.*

1.4.8 Length of a module: For a module M of finite length, the length of any of its composition series (which is independent of the series) is called the *length of the module* and is denoted by $\ell_R(M)$ or simply $\ell(M)$ if there is no confusion about the base ring R.

We define $\ell_R(M) = \infty$ if M has *no composition series*, justifying the fact that such a module is *not a module of finite length*. Note then that $\ell_R(M) = 0 \iff M = (0)$.

1.4.9 Remarks: (i) $\ell(M) \ge 0$ and equality holds $\iff M = (0)$,
(ii) $\ell(M) = 1 \iff M$ is simple and **(iii)** for a non–zero module M, $\ell(M)$ is a measure of departure of M from being simple.

1.4.10 Corollaries: 1. *Let N be a submodule of a module M of finite length. Then $\ell(M) = \ell(N) + \ell(M/N)$ and in particular, $\ell(M) \ge \ell(N)$ with equality $\iff M = N$.*
2. *Sum of finitely many submodules of finite lengths is a module of finite length and $\ell(\sum_{i=1}^{n} M_i) \le \sum_{i=1}^{n} \ell(M_i)$ with equality if and only if*

the sum is a direct sum (1.1.14), i.e.,

$$\ell(\sum_{i=1}^{n} M_i) = \sum_{i=1}^{n} \ell(M_i) \Leftrightarrow \sum_{i=1}^{n} M_i = \bigoplus_{i=1}^{n} M_i$$

3. *If a vector space has some finite basis, then any other basis is finite and all bases have the same number of elements equal to $\ell(V)$, called the dimension of V.*

1.4.11 Corollary: *For any vector space V over a division ring D, the ring $R = \text{End}_D(V)$ is a module of finite length (as an R-module) if and only if V is finite dimensional over D.*

1.4.12 Remark: If $\dim_D(V) = r$, then $R = \text{End}_D(V)$ is a module of finite length as a module over D as well as over itself. Its lengths are given by $\ell_R(R) = r$ whereas $\ell_D(R) = r^2$. The second follows simply because $\dim_D(R) = \dim_D(M_r(D)) = r^2$. The first follows because R has a composition series, namely, $R = R_0 \supset R_1 \supset \cdots \supset R_r = (0)$, where R_i is the left ideal consisting of all matrices whose first i columns are zero, $0 \le i \le r$. By Ex.(1.11.39) below, we note that the simple quotients R_i/R_{i+1} are all mutually isomorphic.   ■

1.5   Artinian Rings

1.5.1 Artinian ring: A ring R is called (*left*) *Artinian* if it is Artinian as a left module over itself, or equivalently, d.c.c or minimum condition holds for left ideals of R.

1.5.2 Examples: Fields, division rings, finite rings are all Artinian. The ring of integers $\mathbb{Z}$ is not Artinian.

1.5.3 *A quotient ring of an Artinian ring is Artinian (whereas a subring need not be Artinian).*

1.5.4 *A finitely generated module over an Artinian ring is Artinian.*

1.5.5 *Matrix rings over Artinian rings (in particular, over division rings), are Artinian.*

1.5.6 *Let R be an Artinian ring. Then we have the following.*

1. *Every non-zero divisor in R is a unit. In particular, an Artinian integral domain is a division ring.*
2. *If R is commutative, every prime ideal is maximal.* ∎

1.6 Noetherian Rings

1.6.1 Noetherian ring: A ring R is called (*left*) *Noetherian* if it is Noetherian as a left module over itself, or equivalently, a.c.c or maximum condition holds for left ideals or every left ideal is finitely generated.

1.6.2 Examples: Fields, division rings, finite rings, principal ideal rings, etc., are all Noetherian. In particular, the ring of integers $\mathbf{Z}$ is Noetherian.

1.6.3 *A quotient ring of a Noetherian ring is Noetherian (whereas a subring need not be Noetherian).*

1.6.4 *A finitely generated module over a Noetherian ring is Noetherian.*

1.6.5 *Matrix rings over Noetherian rings, (in particular, over division rings), are Noetherian.*

1.6.6 Hilbert Basis Theorem: *A ring R is Noetherian $\Leftrightarrow R[X]$ is Noetherian. Consequently, R is Noetherian $\Leftrightarrow R[X_1, X_2, \cdots, X_n]$ is Noetherian for any finitely many variables. Moreover, a finitely generated ring over a Noetherian ring, generated by a commuting set of generators, is Noetherian.*

1.6.7 *Let V be a vector space over a division ring D and R = $\mathrm{End}_D(V)$. Then the following are equivalent.*
(i) *R is Artinian,* **(ii)** *V is finite dimensional over D and*
(iii) *R is Noetherian.* See (2.1.9) below, for a similar statement in a more general setup. ∎

1.7   Jacobson Radical

1.7.0   Radical ideal: A two–sided ideal I in a ring R is called a *radical ideal* with respect to a specified property $\mathcal{P}$ if
1. the ideal I possesses the property $\mathcal{P}$ and
2. the ideal I is maximal for the property $\mathcal{P}$, i.e., if J is a 2–sided ideal of R having the property $\mathcal{P}$, then $J \subseteq I$.

It is obvious by (2) that a radical ideal is unique if it exists.

Note that there are several kinds of radical ideals defined and studied in a ring in various contexts. Notable among them are the *nil radical* and the *Jacobson radical*. We shall be mainly interested in the Jacobson radical $J(R)$. The existence of the nil radical $N(R)$ is given in Ex.(1.11.32) and then by Ex.(1.11.33) we have $N(R) \subseteq J(R)$.

1.7.1   Jacobson radical: The *Jacobson radical* of a ring R is defined as the radical ideal of R with respect to the property that "A 2–sided ideal I is such that $1 - a$ is a unit in R for all $a \in I$" and it is denoted by $J(R)$. In other words, $J(R)$ is the largest 2–sided ideal of R such that $1 - a$ is a unit for all $a \in J(R)$.

Before we see the existence of the Jacobson radical, let us note the following special cases.

1.7.2   Examples: **1.** $J(\mathbb{Z}) = (0)$ and $J(R) = (0)$ for any field or a division ring R.
2. $J(M_r(D)) = (0)$ for a division ring D since $M_r(D)$ has no 2–sided ideals other than (0) and $M_r(D)$ and the latter cannot be a candidate.
3. If R is a commutative local-ring with its unique maximal ideal M, then obviously $J(R) = M$

1.7.3   Left Jacobson radical: For any ring R, the *intersection of all maximal left ideals* of R is called the *left Jacobson radical* or, simply the *left radical* of R and is denoted by $J_\ell(R)$.

1.7.4   Examples: **1.** The left radical of a division ring is (0).
2. The radical of $\mathbb{Z}$ is (0).
3. The radical of a local ring is its unique maximal ideal.

4. The radical of $\mathbb{Z}/n\mathbb{Z}$ is $m\mathbb{Z}/n\mathbb{Z}$ where m is the product of all distinct prime divisors of n. For instance, $J_\ell(\mathbb{Z}/36\mathbb{Z}) = 6\mathbb{Z}/36\mathbb{Z}$, $J_\ell(\mathbb{Z}/64\mathbb{Z}) = 2\mathbb{Z}/64\mathbb{Z}$ and $J_\ell(\mathbb{Z}/180\mathbb{Z}) = 30\mathbb{Z}/180\mathbb{Z}$.

5. The left radical of $M_n(D)$ is (0) for any division ring D.

1.7.5 Theorem: *For any ring R, its left radical $J_\ell(R)$ is the intersection of the annihilators of all simple left modules over R. In particular, $J_\ell(R)$ is a 2–sided ideal of R.*

1.7.6 Corollary: *Given a ring R with its left radical $J_\ell(R)$, the left radical of the quotient $R/J_\ell(R)$ is zero, i.e., $J_\ell(R/J_\ell(R)) = (0)$.*

1.7.7 Theorem: $J_\ell(R) = \{x \in R \mid 1 - yx \text{ is a unit, } \forall\, y \in R\}.$

1.7.8 Theorem: *$J_\ell(R)$ is the largest left ideal of R such that $1 - a$ is a unit for every $a \in J_\ell(R)$.*

1.7.9 Right Jacobson radical: The *intersection of all maximal right ideals* of R is called the *right Jacobson radical* or simply the *right radical* of R and is denoted by $J_r(R)$.

1.7.10 Remarks: Proceeding as above, we can prove that $J_r(R)$ has the following properties.

1. $J_r(R)$ is a 2–sided ideal of R.
2. $J_r(R) = \{x \in R \mid 1 - xy \text{ is a unit, } \forall\, y \in R\}.$
3. $J_r(R)$ is the largest right ideal of R such that $1 - b$ is a unit for all $b \in J_r(R)$.

1.7.11 Theorem: *For any ring R, the left and right Jacobson radicals coincide and the 2–sided ideal $J(R) = J_\ell(R) = J_r(R)$ is the Jacobson radical of R. In particular, the Jacobson radical of a local ring is its (unique) maximal ideal.*

1.7.12 Nakayama Lemma: *If M is a finitely generated module over a ring R such that $J(R)M = M$, then $M = (0)$.*

Recall that for any subset A of R, the set AM stands for the submodule of M generated by $\{ax \mid a \in A \text{ and } x \in M\}$. It is important to note that the assumption that M is finitely generated is necessary in the Nakayama Lemma. (Give an example.) ∎

1.8   Radical of an Artinian Ring

The following are some standard facts related to the Jacobson radical of an Artinian ring.

1.8.1   *The Jacobson radical of an Artinian ring is the intersection of some finitely many maximal left (resp., right) ideals.*

1.8.2   *The Jacobson radical of an Artinian ring R is nilpotent. In fact, $J(R)$ is the largest nilpotent (left or right or 2-sided) ideal of R and consequently $N(R) = J(R)$.*

1.8.3   *In an Artinian ring, every nil ideal is nilpotent (since such an ideal is contained in the radical which is nilpotent).*

1.8.4   *There are only finitely many maximal ideals in a commutative Artinian ring, i.e., it is a semi-local ring.*

1.8.5 Remark: We have seen examples of Artinian modules which are not Noetherian and vice-versa and some which are neither. On the other hand, there are Noetherian rings which are not Artinian and some which are neither. Nevertheless, it is a remarkable fact that "*Every Artinian ring is Noetherian*" as we shall prove in (2.4.9) below. The proof in the case of *commutative* Artinian ring, however, is relatively easy, as sketched below.

1.8.6   *A commutative ring is Artinian if and only if it is Noetherian in which every prime ideal is maximal.*
1. If m_i, $1 \le i \le r$, are all the maximal ideals of a commutative semi-local ring R, then $J = J(R) = m_1 \cap \cdots \cap m_r = m_1 \cdots m_r$ is the product of all the maximal ideals and hence $J^\ell = m_1^\ell \cdots m_r^\ell$.
2. If $J^\ell = (0)$, then (by Ex.(1.11.19) below) R is isomorphic to the product of the local rings $\prod_{i=1}^r R/\, m_1^\ell$ each of whose maximal ideal is nilpotent.
3. A commutative local ring whose maximal ideal is nilpotent is Artinian if and only if it is Noetherian (Ex.(1.11.29) below).
4. In a Noetherian ring R, every ideal contains a finite product of prime ideals and hence the ideal $(0) = P_1 \cdots P_k$ is a product of prime ideals. If each prime ideal is maximal, then R is semi-local (with its

maximal ideals being the distinct ones among the P_i's). Consequently, $J(R) = N(R)$ which is nilpotent. ∎

1.9 Algebras

1.9.1 Algebra: By an *algebra* A over a commutative ring R (with 1), we mean a ring A which is also a unitary R–module such that $a(xy) = (ax)y = x(ay)$ for all $a \in R$ and $x, y \in A$.

Subalgebras are defined to be subrings which are also submodules.

1.9.2 *A ring A is an R-algebra $\Longleftrightarrow$ there exists a unitary homomorphism of rings $\lambda : R \to A$ such that $\lambda(R) \subseteq$ Centre(A).*

1.9.3 Examples: 1. *Any ring is an algebra over $\mathbb{Z}$.*
2. *A ring is an algebra over any of its central subrings.*

1.9.4 Quotient algebra: Given a 2–sided ideal I of an R–algebra A, the quotient ring A/I is an R–algebra in a natural way and is called the *quotient algebra* of A modulo I.

1.9.5 Homomorphism of algebras: Given R–algebras A and B, by an *R–algebra homomorphism* $\lambda : A \to B$, we mean a *homomorphism of rings* which is also a *homomorphism of R-modules*.

The set of all homomorphisms of R–algebras A and B is denoted by $\text{Hom}_{R-\text{alg}}(A, B)$. Thus we have

$$\text{Hom}_{R-\text{alg}}(A, B) = \text{Hom}_{\text{rings}}(A, B) \bigcap \text{Hom}_{R-\text{mod}}(A, B),$$

the intersection being in $\text{Hom}_{\mathbb{Z}}(A, B)$. The notions of monomorphism, epimorphism, isomorphism, kernel of a homomorphism, etc., are exactly as for rings or modules. The homomorphism theorems are identical with their counter parts in rings. For the sake of ready reference, we give the statements.

1.9.6 Epimorphism Theorem: *Suppose $f : A \to B$ is an epimorphism of R-algebras with $I = \text{Ker } f$. Then there exists a unique isomorphism $\widetilde{f} : A/I \to B$ such that $f = \widetilde{f} \circ \eta$, where η is the natural* *map given by* $\eta: A \longrightarrow A/I, \quad x \mapsto x + I$, *i.e., the following diagram is commutative.*

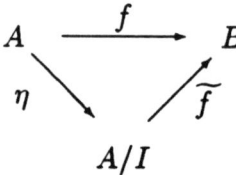

1.9.7 Theorem (Quotient of a quotient): *Suppose* $I \subseteq J \subseteq A$ *are 2-sided ideals of an* R-*algebra* A. *Then there exists a natural isomorphism* $\widetilde{\eta} : (A/I)/(J/I) \xrightarrow{\sim} A/J$, *making the following diagram commutative:*

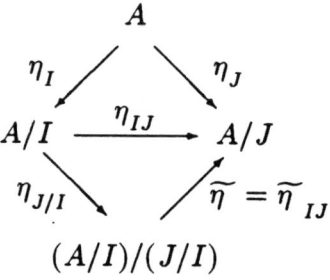

where $\eta_{IJ} : x + I \mapsto x + J$ *for all* $x \in A$.

The following statement for modules has no parallel for rings or algebras.

1.9.8   Theorem (Quotient of a sum): *Suppose* P, N *are sub-modules of an* R-*module* M. *Then there exist natural isomorphisms*

(i) $\dfrac{P + N}{P} \approx \dfrac{N}{N \cap P}$ *and*   (ii) $\dfrac{P + N}{N} \approx \dfrac{P}{P \cap N}.$ ■

1.10   Tensor Products

1.10.1 Balanced map: Given a right R-module M, a left R-module N and an abelian group P, by a *balanced map* $f : M \times N \rightarrow P$, we mean the following.

1. f is *biadditive*, i.e.,

$$\begin{cases} \text{(a)} & f(x + x', y) = f(x, y) + f(x', y), \ \forall \ x, x' \in M, \ y \in N, \\ \text{(b)} & f(x, y + y') = f(x, y) + f(x, y'), \ \forall \ x \in M, \ y, y' \in N. \end{cases}$$

2. $f(xa, y) = f(x, ay), \ \forall \ x \in M, \ y \in N$ and $a \in R$.

The set of all balanced maps from $M \times N$ to P is denoted by $\mathrm{Bal}_R(M, N; P)$.

1.10.2   Tensor product: Given a right R–module M and a left R–module N, by a *tensor product* of M and N over R, we mean a pair (T, φ), where T is an *abelian group* and $\varphi : M \times N \to T$ is a *balanced map* satisfying the following *universal property*:

Given an abelian group P and a balanced map $f : M \times N \to P$, there exists a unique homomorphism $\widetilde{f} : T \to P$ of abelian groups such that $f = \widetilde{f} \circ \varphi$, i.e., the following diagram is commutative.

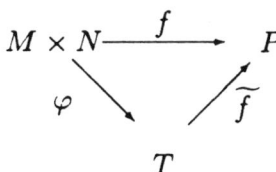

The universal property of tensor product implies that the map, $f \mapsto \widetilde{f}$, is a natural bijection of the set $\mathrm{Bal}_R(M, N; P)$ onto $\mathrm{Hom}_{\mathbb{Z}}(T, P)$ for every abelian group P.

1.10.3   Theorem: *Tensor product of modules exists and is unique upto isomrphism.*

Uniqueness is a trivial consequence of the universal property stated above. As for the existence, let F be the *free abelian group* (i.e., a free module over $\mathbb{Z}$) with the set $M \times N$ *as a basis* and L be the subgroup of F spanned by the following three classes of elements in F:

$$\begin{cases} \text{(a)} & (x + x', y) - (x, y) - (x', y), \ \forall \ x, x' \in M, \ y \in N, \\ \text{(b)} & (x, y + y') - (x, y) - (x, y'), \ \forall \ x \in M, \ y, y' \in N \text{ and} \\ \text{(c)} & (xa, y) - (x, ay), \ \forall \ x \in M, \ y \in N \text{ and } a \in R. \end{cases}$$

It is easy to check that (T, φ) is a tensor product of M and N where $T = F/L$ and $\varphi : M \times N \rightarrow T$ is the natural map $(x, y) \mapsto (x, y) + L$ for all $(x, y) \in M \times N$. $\diamond$

Notation: 1. The tensor product of M and N over R is denoted by $(M \otimes_R N, \otimes_R)$ or simply by $M \otimes_R N$, where
2. the balanced map φ is denoted by $\otimes_R$ or simply by $\otimes$ and
3. for $x \in M$ and $y \in N$, the element $\varphi(x, y) = x \otimes y$ is called a *decomposable tensor*.

1.10.4 Remarks: 1. It is easy to see that the set of decomposable tensors span T, i.e., every element $z \in T$ can be written as a finite sum of the form $z = \sum_{i=1}^{r} x_i \otimes y_i$ for suitable $x_i \in M$ and $y_i \in N$.
2. Decomposable tensors satisfy the following relations.
$$\begin{cases} \text{(a)} & (x + x') \otimes y = x \otimes y + x' \otimes y, \ \forall \ x, x' \in M, \ y \in N, \\ \text{(b)} & x \otimes (y + y') = x \otimes y + x \otimes y', \ \forall \ x \in M, \ y, y' \in N \text{ and} \\ \text{(c)} & (xa) \otimes y = x \otimes (ay), \ \forall \ x \in M, \ y \in N \text{ and } a \in R. \end{cases}$$
3. Tensor product of non–zero modules could be zero.
4. We emphasize the fact that the tensor product is *only an abelian group* and <u>not</u> an R–module, in general. However, it is an R–module if R is commutative, (see (1.10.11)(2) below).

1.10.5 Theorem: *Tensor product commutes with direct sums*, i.e.,
$$M \otimes_R \left(\bigoplus_{i \in I} N_i \right) \simeq \bigoplus_{i \in I} (M \otimes_R N_i).$$
Proof: The idea being the same in the general case, we indicate a proof for the case of $M \otimes_R (P \oplus Q)$:

Consider the map $\lambda : M \times (P \oplus Q) \rightarrow (M \otimes_R P) \oplus (M \otimes_R Q), (x, y+z) \mapsto x \otimes y + x \otimes z$, which is obviously balanced and hence gives rise to an additive homomorphism $\tilde{\lambda} : M \otimes_R (P \oplus Q) \rightarrow (M \otimes_R P) \oplus (M \otimes_R Q)$. On the other hand, the map $\mu_P : M \times P \rightarrow M \otimes_R (P \oplus Q), (x, y) \mapsto x \otimes y$, is balanced giving rise to the additive homomorphism $\widetilde{\mu_P} : M \otimes_R P \rightarrow M \otimes_R (P \oplus Q)$. Similarly, we get an additive homomorphism $\widetilde{\mu_Q} : M \otimes_R Q \rightarrow M \otimes_R (P \oplus Q)$. Putting together we get an additive homomorphism
$$\tilde{\mu} = \widetilde{\mu_P} + \widetilde{\mu_Q} : (M \otimes_R P) \oplus (M \otimes_R Q) \longrightarrow M \otimes_R (P \oplus Q).$$

It is obvious that $\tilde{\lambda}$ and $\tilde{\mu}$ are inverse to each other, as required. $\Diamond$

1.10.6 Tensor product of homomorphisms: Let M, P be right R–modules and N, Q be left R–modules. Let $f \in \mathrm{Hom}_R(M, P)$ and $g \in \mathrm{Hom}_R(N, Q)$. Then there exists an additive homomorphism, $x \otimes y \mapsto f(x) \otimes g(y)$, from $M \otimes N$ to $P \otimes Q$, called the *tensor product of homomorphisms*, and is denoted by $f \otimes g$.

1.10.7 Remarks: The tensor products of homomorphisms have the following properties.
1. *Compositions:* $(f \otimes g) \circ (\lambda \otimes \mu) = (f \circ \lambda) \otimes (g \circ \mu)$.
2. *Epimorphisms:* If f and g are both epimorphisms, so is $f \otimes g$.
3. *Isomorphisms:* If f and g are both isomorphisms, so is $f \otimes g$; in fact, we have $(f \otimes g)^{-1} = f^{-1} \otimes g^{-1}$.
4. However, if f and g are both monomorphisms, (even if one of them is an isomorphism), $f \otimes g$ need not be a monomorphism.

1.10.8 Bimodule: Let R and S be two rings and M be an R–module as well as an S–module. Then M is called an (R, S)–*bimodule* if the scalar multiplications on M commute, i.e., $a(bx) = b(ax)$ for all $a \in R$, $x \in M$ and $b \in S$.

Note: 1. In case M is a left R–module and a right S–module, then M is an (R, S)–bimodule means that $a(xb) = (ax)b$.
2. An (R, R)–bimodule is simply called an R–bimodule .

1.10.9 Remarks: 1. Any R–module M is a $(\mathbb{Z}, R)$–bimodule.
2. A module M over a commutative ring R is naturally an R–bimodule by defining the right module structure to be the same as the left module, i.e., $xa = ax$ for all $a \in R$ and $x \in M$.
3. A ring R, considered as a (left, right)–module over any of its subrings, (under left and right multiplications), is a bimodule.

1.10.10 Proposition: *Let M be a (left, right)–bimodule and N a left module (over R). Then the abelian group $M \otimes_R N$ has a natural structure of a left R–module under the scalar multiplication that*

$$a \cdot (x \otimes y) = (ax) \otimes y, \; \forall \, a \in R, x \in M \text{ and } y \in N.$$

Proof: For a fixed $a \in R$, the map $\lambda_a : M \times N \to M \otimes_R N$, $(x, y) \mapsto$

$(ax) \otimes y$ is balanced and hence gives rise to an additive endomorphism $\widetilde{\lambda}_a$ of $M \otimes_R N$. It can be seen that the map $\widetilde{\lambda} : R \to \mathrm{End}_{\mathbb{Z}}(M \otimes_R N)$, $a \mapsto \widetilde{\lambda}_a$, is a homomorphism of rings, as required. $\Diamond$

1.10.11 Corollaries: 1. *If M is a right module and N is a (left, right)-bimodule, then $M \otimes_R N$ is a right R-module.*
2. *Let M and N be modules over a* commutative *ring R. Then $M \otimes_R N$ is an R-module via the bimodule M or N and these two module structures coincide. Furthermore, we have*

$$a(x \otimes y) = (ax) \otimes y = (xa) \otimes y = x \otimes (ay), \ \forall \, x \in M, \, y \in N \text{ and } a \in R.$$

1.10.12 Proposition (Associativity of tensor products): *Let R and S be rings. Suppose L is a right R-module, M is an (R, S)-bimodule and N is a left S-module. Then the tensor products*

$$L \otimes_R (M \otimes_S N) \quad and \quad (L \otimes_R M) \otimes_S N$$

are defined and are isomorphic to each other (as abelian groups).

1.10.13 Base change: Let M be a left R-module and S a ring containing R as a subring. Considering S as an (S, R)-bimodule (i.e., a left S-module and a right R-module), the tensor product $S \otimes_R M$ becomes a left S-module, called the *base change* or the *extension of scalars* from R to S.

1.10.14 Transitivity of base change: Let $R \subseteq S \subseteq T$ be a *tower of rings* and M be a left R-module. Then the natural T-linear maps $(\forall \, t \in T, a \in S$ and $x \in M)$

$$T \otimes_R M \longrightarrow T \otimes_S (S \otimes_R M), \quad t \otimes x \mapsto t \otimes (1 \otimes x) \text{ and}$$
$$T \otimes_S (S \otimes_R M) \longrightarrow T \otimes_R M, \quad t \otimes (a \otimes x) \mapsto ta \otimes x,$$

are inverse to each other. In other words, *base change is transitive.*

1.10.15 Proposition: 1. *For any left R-module M, the natural map $R \otimes_R M \to M$, $a \otimes x \mapsto ax$, is an isomorphism of left R-modules.*
2. *Base change of a free module is free. In fact, if X is an R-basis of M, then $\{1 \otimes x | x \in X\}$ is an S-basis of $S \otimes_R M$.*

3. *Tensor product of free modules is free; in fact, if $\{x_i | i \in I\}$ and $\{y_j | j \in J\}$ are bases for M and N respectively, then $\{x_i \otimes y_j | (i,j) \in I \times J\}$ is a basis for $M \otimes N$.*

4. *In paricular, we have $dim_K(V \otimes_K W) = (\dim_K V)(\dim_K W)$ for all (finite dimensional) vector spaces V and W over a field K.*

1.10.16 Tensor product over commutative rings: Let L, M and N be modules over a commutative ring R. Then we have the following.

1. Commutativity of tensor product: The R–modules $M \otimes_R N$ and $N \otimes_R M$ (1.10.11)(2) are naturally isomorphic to each other under the map $x \otimes y \mapsto y \otimes x$.

2. Associativity of tensor product: The R–modules $(L \otimes_R M) \otimes_R N$ and $L \otimes_R (M \otimes_R N)$ are naturally isomorphic to each other under the map $(x \otimes y) \otimes z \mapsto x \otimes (y \otimes z)$.

1.10.17 Tensor product of algebras: Let A and B be algebras over a commutative ring R. Then the R–module $A \otimes_R B$ has a natural structure of an R–algebra under the multiplication defined (on decomposable elements) by $(a \otimes b)(x \otimes y) = ax \otimes by$ for all $a, x \in A$ and $b, y \in B$.

1.10.18 Proposition: *It is easy to see the following.*

1. *For all $a \in A$ and $b \in B$, the elements $a \otimes 1$ and $1 \otimes b$ commute with each other in $A \otimes_R B$ and the ring $A \otimes_R B$ is generated by such elements. In particular, if A and B are commutative, so is $A \otimes_R B$.*

2. *If A is generated by X and B is generated by Y as R–algebras, then $A \otimes_R B$ is generated by $\{x \otimes 1 | x \in X\} \cup \{1 \otimes y | y \in Y\}$ as an R–algebra. In paricular, if A and B are finitely generated over R, so is $A \otimes_R B$.*

3. *Tensor product of polynomial algebras is a polynomial algebra, i.e., we have an R–isomorphism of polynomial rings,*

$$(R[X_i : i \in I]) \otimes_R (R[Y_j : j \in J]) \simeq R[X_i \otimes 1, 1 \otimes Y_j : i \in I, j \in J].$$

4. *The natural maps $\eta_A : A \to A \otimes_R B$ $(a \mapsto a \otimes 1)$ and $\eta_B : B \to A \otimes_R B$ $(b \mapsto 1 \otimes b)$ are R–algebra homomorphisms whose images commute with each other and generate $A \otimes_R B$.*

5. *The maps η_A and η_B are neither injective nor surjective.*

6. *The maps η_A and η_B satisfy the* universal property *that*
(i) given an R–algebra C and
(ii) *R–algebra homomorphisms $\varphi_A : A \rightarrow C$ and $\varphi_B : B \rightarrow C$*
such that
(iii) *$\varphi_A(a)\varphi_B(b) = \varphi_B(b)\varphi_A(a)$, $\forall\, a \in A$ and $b \in B$, then*
(iv) *there exists a* unique *R–algebra homomorphism $\varphi\colon A \otimes_R B \rightarrow C$*
such that $\varphi_A = \varphi \circ \eta_A$ and $\varphi_B = \varphi \circ \eta_B$, i.e., the diagram

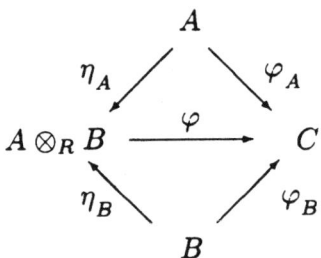

is commutative. In fact, we have $\varphi(a \otimes b) = \varphi_A(a)\varphi_B(b)$ for all $a \in A$
and $b \in B$. Consequently, for all commutative R–algebras C, we have

$$\mathrm{Hom}_{R\text{--alg}}(A \otimes_R B, C) = \mathrm{Hom}_{R\text{--alg}}(A, C) \times \mathrm{Hom}_{R\text{--alg}}(B, C).$$

7. *If I and J are 2–sided ideals in A and B respectively, then we have*
a natural isomorphism of R–algebras

$$(A/I) \otimes_R (B/J) \simeq (A \otimes_R B)/(I \otimes B + A \otimes J).\qquad\blacksquare$$

1.11 Exercises

Unless otherwise stated explicitly, a ring means a ring with unity, whether or not commutative. When two or more ideals are referred to, all are supposed to be ideals of the same kind (left/right/2–sided). All modules are assumed unitary. A homomorphism of rings is not necessarily unitary.

1. Let $R = M_n(\mathbf{Z})$ with $n \geq 2$. Give examples of matrices in R having the following properties. (i) $A \in R$ is such that A is a *zero–divisor* or a *divisor of zero* (i.e., $AX = 0$ or $YA = 0$ for some *non–zero* X or non–zero Y in R) but <u>neither</u> *nilpotent* (i.e., $A^m \neq 0$ for all $m \in \mathbf{N}$) <u>nor</u> an *idempotent* (i.e., $A^2 \neq A$).
   (ii) $A, B \in R$ are such that both A and B are idempotents but $A + B$ is *not* an idempotent. Verify that for such a pair, $AB + BA \neq 0$.
   (iii) $A, B \in R$ are such that both A and B are nilpotent but $A + B$ is *not* nilpotent. Verify that for such a pair $AB \neq mBA$, $\forall \, m \in \mathbf{Z}$.

2. Let a, b be two commuting elements in a ring R, i.e., $ab = ba$. For any positive integer n, prove the binomial expansion that

$$(a + b)^n = \sum_{i=0}^{n} \binom{n}{i} a^{n-i} b^i.$$

   Hence or otherwise, show that the set of all nilpotent elements in a commutative ring is an ideal.

3. Let R be a commutative ring whose characteristic is a prime p. Then show that $(a + b)^p = a^p + b^p$ for all a, b in R. Hence deduce that

$$(a + b)^{p^n} = a^{p^n} + b^{p^n}$$

   for all a, b in R and $n \in \mathbf{N}$. Show that this result need not be true if the characteristic is *not* a prime (even if it is a power of a prime).

4. Show that the characteristic of a simple ring is either 0 or a prime.

5. Let $Z = Z(R)$ be the centre of a ring R. Show that
   (i) $Z[X]$ is the centre of $R[X]$ but
   (ii) $M_n(Z)$ is *never* the centre of $M_n(R)$ if $n \geq 2$ and
   (iii) determine the centre of $M_n(R)$ for all $n \in \mathbf{N}$.

6. Given any ring R and a positive integer n, show that the map *transpose* : $M_n(R^0) \to \left(M_n(R)\right)^0$, sending a matrix to its transpose, is an isomorphism (where R^0 is the ring opposite to R (1.1.1)).

7. For an ideal I of a ring R, let $J = M_n(I) \subseteq S = M_n(R)$. Show that
   (i) J is an ideal in S·of the same type as that of I in R,
   (ii) every 2–sided ideal of S is of the form $M_n(I)$ for a unique 2–sided ideal I of R but (iii) there are left/right ideals of S which are *not* of the form $M_n(I)$ if $n \geq 2$.

8. Show that the *annihilator* of an element $x \in M$ of an R–module M, i.e., $\mathrm{Ann}_R(x) = \{a \in R | ax = 0\}$, is a left ideal but need not be 2–sided. However, show that the annihilator of a module, i.e.,

$$\mathrm{Ann}_R(M) = \{a \in R \mid ax = 0,\ \forall\ x \in M\} = \bigcap_{x \in M} \mathrm{Ann}_R(x)$$

   is a 2–sided ideal, called the *annihilator ideal* of the module.

9. Show that the annihilator ideals of isomorphic modules are *equal* (not just isomorphic) but modules whose annihilators are equal need not be isomorphic.

10. Let R be a ring and $S = R \times R$. Show that
    (i)   both $I = R \times \{0\}$ and $J = \{0\} \times R$ are 2–sided ideals of S
    (ii) each is generated by a *central idempotent* (i.e., an idempotent which is in the centre of S),
    (iii) $S/I \simeq R \simeq S/J$ as R–modules but *not* as S–modules and
    (iv) consequently, I and J are *not* isomorphic as S–modules.

11. Let a, b be a pair of *orthogonal idempotents* (i.e., $a^2 = a$, $b^2 = b$ and $ab = 0 = ba$) in a ring R. Show that the principal left/right ideals generated by a and b are such that $(a)_\ell \cap (b)_\ell = (0) = (a)_r \cap (b)_r$. However, give examples so that the principal 2–sided ideals are such that $(a) \cap (b) \neq (0)$.

12. Show that a left ideal I of a ring R is a direct summand in R if and only if I is generated by an idempotent. Deduce that a non–zero ideal in a local ring is not a direct summand.

13. Give an example to show that if an ideal I in a commutative ring R is finitely generated, its *radical* $\sqrt{I} = \{a \in R \mid a^n \in I \text{ for some } n \in \mathbf{N}\}$, need not be finitely generated.

14. Show that a finitely generated left ideal I is a direct summand if $I^2 = I$. Is the finite generation essential?

15. Let e be a non–zero idempotent in a ring R. Show that the principal left ideal $(e)_\ell$ is minimal (resp. maximal) if and only if the principal right ideal $(e)_r$ is so.

16. Let e and f be two orthogonal idempotents in a ring R. Show that the principal left (resp. right) ideal generated by $e + f$ is the direct sum of the principal left (resp. right) ideals generated by e and f.

17. A non–zero idempotent e in a ring R is called *primitive* if the left ideal Re is minimal. Show that a non–zero idempotent e is primitive if and only if e is not a sum of (non–zero) orthogonal idempotents.

18. If I and J are ideals in a commutative ring R which are *coprime* or *comaximal*, i.e., $I + J = R$, show that I^m and J^n are coprime for all $m, n \in \mathbf{N}$. (Hint: If $K = I^m + J^n \neq R$, apply Zorn's lemma to get a maximal ideal containing K.)

19. (**Chinese Remainder Theorem**): Let R be a commutative ring and I and J be ideals coprime to each other. Show that
    (i) $I \cap J = IJ$ and (ii) the natural map $f : R/IJ \rightarrow (R/I) \times (R/J)$, sending $x + IJ$ to $(x + I, x + J)$, for all $x \in R$, is well–defined and an isomorphism of rings.

20. Show that a finite sum of finitely generated modules is finitely generated and that a quotient of a finitely generated module is finitely generated. However, give an example of two submodules P and Q of a module M such that both P and $P + Q$ are finitely generated but Q is not. Verify that for such a pair, $P \cap Q \neq (0)$.

21. Give examples to show that maximal or minimal submodules need not exist in a non–zero module.

22. Show that an R–module M is cyclic i.e., generated by one element, if and only if $M \simeq R/I$ for some left ideal I of R. Deduce that M is simple if and only if I is a maximal left ideal. However, give an example of a maximal 2–sided ideal I such that R/I is not a simple R–module.

23. Show that a simple R–module M is free if and only if R is a division ring and M is of dimension 1 over R.

24. Let P and Q be submodules of a module M such that both M/P and M/Q are Artinian (resp. Noetherian). Show that $M/(P \cap Q)$ and $M/(P + Q)$ are Artinian (resp. Noetherian). (Hint: Realise the first as a submodule and the other as a quotient of $M/P \times M/Q$.)

25. Let M be a Noetherian R–module with its *annihilator ideal* $I = \text{Ann}_R(M)$. Show that R/I is a Noetherian ring. (Hint: Realise R/I as a submodule of M^n if M is generated by n elements.) Give an example to show that this need not be true for the Artinian case.

26. Show that a monomorphism $f \in \text{End}_R(M)$ of an Artinian module M is an automorphism. (Hint: Look at the descending chain of submodules $f^n(M), n \in \mathbf{N}$.) Give an example to show that it is not true for the Noetherian case.

27. Given left R–modules M and N, show that the additive group $\text{Hom}_R(M, N)$ need not be an R–module. However, show that it is a left R–module (resp. right R–module) if M (resp. N) is an R–bimodule (1.10.8).

28. Let R be commutative and M and N be free and finitely generated R–modules. Show that $\text{Hom}_R(M, N)$ is a free R–module. Is the finite generation assumption essential for both M and N?

29. Give an example of a finitely generated 2–sided ideal I in a ring R such that R/I is Artinian (resp. Noetherian) but R is not. However, show that R is Artinian (resp. Noetherian) if I is *nilpotent*. (Hint: For each i, I^i/I^{i+1} is Artinian (resp. Noetherian).)

30. Give an example of a ring R and an element which has a left inverse but not a right inverse. (Try $\text{End}_D(V)$ for an infinite dimensional vector space V over a division ring D.) However, show that such a thing is not possible in a left Noetherian ring. (Hint: If $ab = 1 \neq ba$, then $\{e_i \mid i \in \mathbf{N}\}$ is an infinite set of pairwise orthogonal idempotents where $e_i = b^{i-1}a^{i-1} - b^i a^i$. Now deduce that $\{R(e_1 + \cdots + e_n) \mid n \in \mathbf{N}\}$ is ascending but not stationary.)

31. Let R be a commutative Noetherian local ring with its maximal ideal M. Show that $I = \cap_{i=1}^{\infty} M^i = (0)$. (Hint: Apply Nakayama to I.)

32. **Nil radical:** The *nil radical* of a ring R is defined to be the radical ideal with respect to the property that "A 2–sided ideal is nil" (1.7.0)

and is denoted by $N(R)$, i.e., $N(R)$ *is the largest 2–sided ideal of R such that every element of $N(R)$ is nilpotent* . Prove the following.

Theorem: *For any ring R, the nil radical $N(R)$ exists and it is characterised by*
$N(R) = \{a \in R \mid \text{the principal 2 – sided ideal } (a) \text{ is a nil ideal}\}$.
In particular, the nil radical of a commutative ring is the set of all the nilpotent elements and furthermore, it is also the intersection of all the *prime ideals*.

33. Show that $N(R) \subseteq J(R)$ for any ring R where $J(R)$ is the Jacobson radical of R.

34. If $J = J(R)$, show that $JS = (0)$ for any simple R–module S, (R commutative or not). (Hint: Apply Nakayama lemma if $JS = S$.)

35. Let N be a submodule of a finitely generated R–module M such that $M = N + JM$. Then show that $N = M$. (This is another version of Nakayama lemma.) (Hint: Apply Nakayama to M/N.)

36. Show that R is Noetherian if and only if $R[[X]]$ is so where

$$R[[X]] = \left\{ \sum_{i=0}^{\infty} a_i X^i \mid a_i \in R, \; \forall \, i \in \mathbf{Z}^+ \right\}$$

is the ring of *formal power series* in the variable X with coefficients in R. (Hint: Imitate one of the standard proofs of Hilbert Basis Theorem replacing "leading coefficient" by "coefficient of lowest degree term".)

37. Let $R = \{ \begin{pmatrix} m & 0 \\ x & y \end{pmatrix} \mid m \in \mathbf{Z}$ and $x, y \in \mathbf{Q}\}$. Show that R is a left Noetherian ring but not right Noetherian.

   (Hint: Show that $I = \{ \begin{pmatrix} 0 & 0 \\ 0 & x \end{pmatrix} \mid x \in \mathbf{Q}\}$ is a minimal left ideal and that $R/I = \{ \begin{pmatrix} m & 0 \\ x & 0 \end{pmatrix} \mid m \in \mathbf{Z}$ and $x \in \mathbf{Q}\}$ is Noetherian. Furthermore, show that $\{I_n, n \in \mathbf{N}\}$ is a non–stationary ascending chain of right ideals where $I_n = \{ \begin{pmatrix} 0 & 0 \\ m/2^n & 0 \end{pmatrix} \mid m \in \mathbf{Z}\}$.)

38. Let $R = \{ \begin{pmatrix} x & y \\ 0 & 0 \end{pmatrix} \mid x, y \in \mathbf{Q}\}$. Show that R is a left Artinian ring but not right Artinian. (Note that R is without 1.)

(Hint: Show that $I = \{\begin{pmatrix} 0 & x \\ 0 & 0 \end{pmatrix} \mid x \in \mathbf{Q}\}$ is a minimal left ideal and that $R/I = \{\begin{pmatrix} x & 0 \\ 0 & 0 \end{pmatrix} \mid x \in \mathbf{Q}\}$ is Artinian. Furthermore, show that $\{I_n, n \in \mathbf{N}\}$ is a non–stationary descending chain of right ideals where $I_n = \{\begin{pmatrix} 0 & 0 \\ 2^n m & 0 \end{pmatrix} \mid m \in \mathbf{Z}\}$.)

39. Let $R = M_n(D)$ where D is a division ring. Show that R is a *ring of finite length* (1.4.12). Furthermore, show that *all* the simple quotients of any composition series for R are isomorphic to the *same* minimal left ideal of R.

40. Let R be a ring of finite length. Let M be a free R–module having a finite basis B. Then show that M is a module of finite length and $\ell_R(M) = |B|\ell_R(R)$ (where $|B|$ is the cardinality of B). Deduce that any other basis C for M also must be finite and $|C| = |B|$, i.e., any two bases of a finitely generated free module over a ring of finite length have the same cardinality.

41. Let M be an R–module of finite length. Suppose that an idempotent $e \in R$ annihilates all the *composition factors* of M. Then show that e annihilates M itself.

42. Let R be a commutative ring and A be an R–algebra which is finitely generated as an R–module. Show that any finitely generated left A–module is also finitely generated as an R–module. What happens if A is merely finitely generated as R–algebra?

43. Let R be commutative Noetherian and A be an R–algebra which is finitely generated as an R–module. Suppose M and N are finitely generated left A–modules. Then show that $\mathrm{Hom}_A(M, N)$ is a finitely generated R–submodule of $\mathrm{Hom}_R(M, N)$. What happens if R is not Noetherian?

44. Give an example to show that the tensor product of non–zero modules need not be non–zero. (Try $(\mathbf{Z}/2\mathbf{Z}) \otimes_{\mathbf{Z}} (\mathbf{Z}/3\mathbf{Z})$.)

45. Let R and S be rings of positive characteristics m and n respectively. Suppose that m and n are coprime. Then show that $R \otimes_{\mathbf{Z}} S = (0)$. In particular, the tensor product of fields of unequal positive characteristics is zero.

46. Suppose F is a free right R–module with a basis B and M is a left R–module. Then show that every element $z \in F \otimes_R M$ can be written as $z = \sum_{b \in B} b \otimes z_b$ for *unique* $z_b \in M$ such that $z_b = 0$ for all but finitely many b's. Hence deduce that that the tensor product of free modules is free (1.10.14)(2).

47. Let V and W be vector spaces over a field K. Show that a *decomposable element* $v \otimes w \neq 0$ for all non–zero vectors $v \in V$ and $w \in W$. Can one replace "vector spaces" by "free modules" over a commutative ring (resp. PID)?

48. Suppose F is a free left R–module with a basis B and S is a ring which is an R–bimodule. Then show that $S \otimes_R F$ is a free left S–module with basis $1 \otimes_R B = \{1 \otimes b \mid b \in B\}$. Hence deduce that the base change of a free module is free (1.10.14)(1).

49. Let R and S be as above. Suppose that any two bases of a finitely generated free S–module have the same cardinality. Then show that the same is true for finitely generated free R–modules. Deduce that finitely generated free modules over commutative rings, rings of finite lengths, etc., have well–defined *dimensions*, usually called the *rank* of the free module.

50. Give an example of a ring R and a finitely generated free R–module which has bases of different cardinalities. (Try $M = R = \mathrm{End}_K(V)$ where V is an infinite dimensional vector space over a field K.)

51. Let R be a commutative ring and A be an R–module. Then show that the following are equivalent. (i) A is an R–algebra.
(ii) There exist R–linear maps $\mu_A : A \otimes_R A \to A$, called the *multiplication* and $\iota_A : R \to A$, called the *unity* such that
Associativity: $\mu_A \circ (1_A \otimes \mu_A) = \mu_A \circ (\mu_A \otimes 1_A)$ where 1_A is the identity map of A and
Existence of unity: $\mu_A(1_A(1) \otimes a) = \mu_A(a \otimes 1_A(1)) = a, \ \forall \ a \in A$.

52. Show that the base change from R to S (1.10.13) of a matrix (resp. polynomial, resp. Laurent polynomial) ring is one such. Is a similar result true for a power series ring?

53. Let A and B be R–algebras with I and J respective 2–sided ideals thereof. Then show that $K = I \otimes_R B + A \otimes_R J$ is a 2–sided ideal of $A \otimes_R B$ and that $(A/I) \otimes_R (B/J) \simeq (A \otimes_R B)/K$ (as R–algebras). Hence

or otherwise, deduce that $(\mathbf{Z}/m\mathbf{Z}) \otimes_{\mathbb{Z}} (\mathbf{Z}/n\mathbf{Z}) = \mathbf{Z}/d\mathbf{Z}$ $(\forall\, m, n \in \mathbf{N})$
where $d = \gcd(m, n)$.

54. Let R and S be rings. Suppose that L is a right R–module, M is an (R, S)–bimodule and N is a right S–module. Then show that there is a natural isomorphism of the abelian groups:

$$\operatorname{Hom}_R(L, \operatorname{Hom}_S(M, N)) \simeq \operatorname{Hom}_S(L \otimes_R M, N).$$

In particular, for a commutative ring R, we have
$\operatorname{Hom}_R(L, M^\star) \simeq (L \otimes_R M)^\star$ and $\operatorname{Hom}_R(M, M^\star) \simeq (M \otimes_R M)^\star$, where $M^\star = \operatorname{Hom}_R(M, R)$ is the *dual* of M, etc.

55. Let R be commutative and M and N be free R–modules with M having a finite basis. Show that the natural map $M^\star \otimes_R N \to \operatorname{Hom}_R(M, N)$, $(\sum_i \lambda_i \otimes y_i) \mapsto \varphi$, where $\varphi(x) = \sum_i \lambda_i(x) y_i$ (for all $\lambda_i \in M^\star$, $x \in M$ and $y_i \in N$) is an isomorphism of free R–modules. In particular, $M^\star \otimes_R M \simeq \operatorname{End}_R(M)$.

56. Let K be a field and R be a finite dimensional K–algebra. Suppose M is a finitely generated left R–module. For any *field extension* L of K (i.e., L is a field containing K as a subfield), let $R^L = L \otimes_K R$ and $M^L = L \otimes_K M$. Then show that (i) M^L is a finitely generated left R^L–module and (ii) $(\operatorname{Hom}_R(M, N))^L \simeq \operatorname{Hom}_{R^L}(M^L, N^L)$ (as vector spaces over L) for all left R–modules N.

57. Let R be commutative and S be a commutative R–algebra. Given a polynomial $f(X) \in R[X]$, show that we have a natural isomorphism of S–algebras $S \otimes_R (R[X]/(f(X)) \simeq S[X]/(f(X))$.

58. Let $f(X) \in \mathbf{Q}[X]$ be an *irreducible polynomial* of degree n. Let $K = \mathbf{Q}[X]/(f(X))$ be the induced *finite dimensional field extension of* $\mathbf{Q}$ (such a K is called a *number field*). Let r and $2s$ be the number of real and complex roots of $f(X)$ respectively so that $n = r + 2s$ with $r, s \in \mathbf{Z}^+$. Show that $K \otimes_{\mathbf{Q}} \mathbf{R} \simeq \mathbf{R}^r \times \mathbf{C}^s$ as $\mathbf{R}$–algebras. In particular, (i) the tensor product of fields (over a field) need not be even an *integral domain* and (ii) the base change of a field to another field (over a field) need not be a field (not even an integral domain).∎

1.12 True/False Statements

Determine which of the following statements are true (T) or false (F) or partially true (PT). Justify your answers by giving a proof if (T) or providing a counter–example if (F)/(PT) and supplying the additional hypothesis needed to make it (T) (along with a proof) if (PT), as the case may be.

1. In the ring $M_n(\mathbf{Z}_m)$, every element is a zero–divisor or a unit.

2. A non–zero element a in a ring R is a unit if the 2–sided ideal (a) equals R.

3. The ring $M_n(\mathbf{Z})$ has infinitely many maximal 2–sided ideals.

4. The ring $M_n(\mathbf{Z})$ has infinitely many minimal 2–sided ideals.

5. The ring $M_n(\mathbf{Q})$ has no minimal left ideals.

6. In a ring R, the union of all maximal left ideals is the same as the union of all maximal right ideals.

7. The 2–sided ideal generated by $\begin{pmatrix} 1 & 0 \\ 0 & 0 \end{pmatrix}$ in $R= M_2(\mathbf{Z})$ is R.

8. The ring $M_n(\mathbf{Q})$ is local only if $n = 1$.

9. Integers m and n are coprime only if the ideals (m) and (n) are so.

10. If R is local with M_R its non–units, then R/M_R is a division ring.

11. A non–zero homomorphism from a simple ring is a monomorphism.

12. A unitary homomorphism from a simple ring is a monomorphism.

13. A non–zero homomorphism between simple rings is an isomorphism.

14. A homomorphism of $M_n(\mathbf{C})$ into any ring is 0 or a monomorphism.

15. A homomorphism of $M_n(\mathbf{Z})$ into any ring is 0 or a monomorphism.

16. A matrix ring over a simple ring is a simple ring.

17. The endomorphism ring of a simple module is a division ring.

18. $\mathrm{End}_R(M)$ is of finite length if M is of finite length.

19. A simple ring R is simple both as left and right R–module.

20. A simple R–module is isomorphic to a minimal left ideal of R.

21. A simple R–module is isomorphic to some quotient of R.

22. Any two bases of a finitely generated free module over $\mathrm{End}_D(V)$ have the same cardinality.

23. Any two bases of a finitely generated free-module over a commutative ring have the same number of elements.

24. The only simple and free modules are one dimensional vector spaces.

25. If a ring R is Artinian, then $R[X]$ is Artinian.

26. The length of a finite Cartesian product of modules of finite length is the product of their lengths.

27. A finitely generated module over $M_n(\mathbb{C})$ is a module of finite length.

28. Minimal submodules exist in a module over an Artinian ring.

29. Minimal submodules exist in a finitely generated non–zero module over a Noetherian ring.

30. Minimal and maximal submodules exist in a module of finite length.

31. In a commutative Artinian ring, a prime ideal is maximal.

32. The length of a finite ring is the number of its non–zero elements.

33. A ring of finite length is a finite ring.

34. The ring $\text{End}_D(V)$ is Noetherian but not Artinian.

35. A matrix ring over a ring of finite length is of finite length.

36. Direct sum of an infinite family of non–zero Noetherian modules is not Noetherian.

37. The Jacobson radical of a Noetherian ring is nilpotent.

38. The Jacobson and nil radicals of an Artinian ring are equal.

39. Elements in $M \otimes_R N$ are of the form $x \otimes y$ for some $x \in M$, $y \in N$.

40. Tensor product of free modules is free.

41. Tensor product of finitely generated modules is finitely generated.

42. Tensor product of simple modules is simple.

43. Tensor product of polynomial algebras over a commutative ring is a polynomial algebra.

44. Tensor product of simple rings is simple.

45. Tensor product of epimorphisms is an epimorphism.

46. Tensor product of monomorphisms is a monomorphism.

47. Tensor product of fields (over a field) is a field.

48. Tensor product of local rings is local.   ∎

Chapter 2

Semi–simple Rings and Brauer Group

In this chapter, we shall study "Semi–simple Modules and Rings" leading to the celebrated "Artin-Molien-Wedderburn structure theorem of semi–simple rings". This is very crucial for the developments in the sequel. Though not required for our later purpose, we shall also present some of the basic concepts of the theory of "Central Simple Algebras" and "Brauer Groups". As usual, unless stated otherwise, a ring will mean a ring with unity (commutative or not) and all modules considered are unitary. (Cf. [23], [31], etc.)

2.1 Semi–simple Modules

We recall (1.1.15) that a submodule P of a module M is called a *direct summand* of M if there is a submodule Q such that $M = P \oplus Q$, i.e., every element $z \in M$ can be written as $z = x + y$ for unique $x \in P$ and $y \in Q$.

2.1.1 Proposition: *A submodule P of a module M is a direct summand of M if and only if there is a* projection pr_P *of M onto P, i.e., pr_P is an endomorhism of M such that*
1. $\mathrm{pr}_P^2 = \mathrm{pr}_P$, *i.e., pr_P is an idempotent and*
2. $\mathrm{pr}_P(M) = P$, *i.e.,* $\mathrm{Image}(\mathrm{pr}_P) = P$.

Proof: Let $M = P \oplus Q$. Take $\mathrm{pr}_P = \eta_Q : M \to M/Q = P$, the natural homomorphism sending $x \mapsto x + Q$, $\forall\, x \in M$, treated as an endomorphism of M in a natural way. Since $\mathrm{pr}_P(x) = x$ for all $x \in P$, we get that $\mathrm{pr}_P^2 = \mathrm{pr}_P$, as required.

Conversely, suppose $f \in \mathrm{End}_R(M)$ is such that $f^2 = f$ and $f(M) = P$. Let $Q = \mathrm{Ker}\, f$. It is easy to see that $M = P \oplus Q$, as required. $\diamond$

2.1.2 Remark: Let $M = P \oplus Q$ with pr_P and pr_Q respective projections onto P and Q. Then it is easy to see the following.
1. pr_P and pr_Q are *orthogonal* to each other in the sense that $\mathrm{pr}_P \circ \mathrm{pr}_Q = \mathrm{pr}_Q \circ \mathrm{pr}_P \equiv 0$ and
2. pr_P and pr_Q give a *partition of* 1, i.e., $\mathrm{id}_M = \mathrm{pr}_P + \mathrm{pr}_Q$.

2.1.4 Proposition: *Let $\{P_i, i \in I\}$ be a family of submodules of M such that each P_i is a direct summand of M with projection pr_i. Then $M = \oplus_{i \in I} P_i$ if and only if*

$$\begin{cases} \text{(i)} & \mathrm{pr}_i\text{'s are mutually orthogonal,} \\ \text{(ii)} & \forall\, x \in M, \mathrm{pr}_i(x) = 0 \text{ for all but finitely many } i \in I \text{ and} \\ \text{(iii)} & \mathrm{pr}_i\text{'s give a partition of 1.} \end{cases}$$

Proof: Natural extension of (2.1.2) above. $\diamond$

Notation: In case $M = \oplus_{i \in I} P_i$ with projections pr_i, then we write $\mathrm{id}_M = \sum_{i \in I} \mathrm{pr}_i$. This sum on the right, which is *infinite* if I is, makes sense as a map in view of the condition (ii) above.

2.1.5 Semi–simple module: A module M is called *semi-simple* if every submodule of M is a direct summand.

Examples: 1. Vector spaces are semi–simple (regardless of their dimensions).
2. Simple modules are semi–simple.

2.1.6 Proposition: 1. *A submodule of a semi–simple module is semi–simple.*
2. *A quotient of a semi–simple module M is isomorphic to a submodule of M and hence semi–simple.*

Proof: (1) Let N be a submodule of a semi–simple module M. Let P be a submodule of N. Then there is a submodule Q of M such that $M = P \oplus Q$. But then it follows that $N = P \oplus (Q \cap N)$, as required.

(2) Let N be a submodule of a semi–simple module M. Then there is a submodule P of M such that $M = N \oplus P$ and so $M/N \approx P$, as required. $\diamond$

2.1.7 Theorem: *The following are equivalent for a module M.*
1. *M is semi–simple.*
2. *M is a sum of a family of simple submodules of M.*
3. *M is a direct sum of a family of simple modules.*

Proof: (1) $\Rightarrow$ (2): Let M be semi–simple. By (2.1.6) (1) above, every submodule of M is semi–simple. Secondly, if $M \neq (0)$, M contains simple submodules. (For, take any $x \in M, x \neq 0$ and consider $N = Rx$ which is finitely generated and non–zero. Hence N has a maximal submodule, say P. Since (i) holds for N, there is a submodule Q of N such that $N = P \oplus Q$ and so $Q \approx N/P$ is a simple submodule of N and hence of M.)

To prove (2), we may assume that $M \neq (0)$. Let $\mathcal{M}$ be the family of all simple submodule of M and consider $N = \sum_{S \in \mathcal{M}} S$. Let, if possible, $N \neq M$. By (1), we can write $M = N \oplus P$ for some submodule P of M. Since $P \neq (0)$, again by (1), P has a simple submodule, say S_0. Now $S_0 \in \mathcal{M}$ and so $S_0 \subseteq N$ which gives $S_0 \subseteq N \cap P = (0)$ which is a contradiction to the fact that $S_0 \neq (0)$. Thus $N = M$, as required.

(2) $\Rightarrow$ (3): Suppose $M = \sum_{i \in I} S_i$, S_i simple. If $I = \emptyset$, we have nothing to prove since $M = (0)$. Let $\mathcal{I}$ be the set of subsets J of I such that the sum $\sum_{j \in J} S_j$ is direct, i.e., $\sum_{j \in J} S_j = \bigoplus_{j \in J} S_j$. Notice that $\mathcal{I} \neq \emptyset$ since $\{i\} \in \mathcal{I}$, $\forall\, i \in I$. Partially order $\mathcal{I}$ by set inclusion.

Claim: $\mathcal{I}$ has a maximal member.

To apply Zorn's lemma to $\mathcal{I}$, let $\mathcal{T}$ be a chain in $\mathcal{I}$. Let $T_0 = \cup_{T \in \mathcal{T}} T$ and look at the sum $\sum_{t \in T_0} S_t$. If this sum is not direct, there exist $t_i \in T_0$ and $x_i \in S_{t_i}, 1 \leq i \leq r$ such that *not all* x_i are zero but

$\sum_{i=1}^{r} x_i = 0$. Since $\mathcal{T}$ is a chain, we can find a $T' \in \mathcal{T}$ such that $t_i \in T'$, $\forall\, i$, $1 \le i \le r$. But then it means that the sum $\sum_{t \in T'} S_t$ is *not direct* because of the non–trivial relation that $\sum_{i=1}^{r} x_i = 0$ in $\sum_{t \in T'} S_t$ which is a contradiction. This proves the claim.

Let $J \in \mathcal{I}$ be a maximal member. We have $N = \sum_{j \in J} S_j = \bigoplus_{j \in J} S_j$ and J is a subset of I maximal for this property. If $N \ne M$, then there exists an $i_0 \in I$ such that $S_{i_0} \not\subset N$, i.e., $S_{i_0} \cap N \ne S_{i_0}$ and hence $S_{i_0} \cap N = (0)$ since S_{i_0} is simple. Thus $i_0 \notin J$ and the sum over the larger subset $J' = J \cup \{i_0\}$, namely, $N + S_{i_0} = (\sum_{j \in J} S_j) + S_{i_0} = \sum_{t \in J'} S_t$ is a direct sum contradicting the maximality of J in $\mathcal{I}$. This proves that $M = \bigoplus_{j \in J} S_j$, as required.

$(3) \Rightarrow (1)$: Let $M = \bigoplus_{i \in I} S_i$ with S_i simple. Let N be any submodule of M. If $N = M$, we are through. Suppose that $N \ne M$, i.e., $S_{i_0} \not\subset N$ for some $i_0 \in I$ and so $S_{i_0} \cap N \ne S_{i_0}$ and hence $S_{i_0} \cap N = (0)$, i.e., $N + S_{i_0} = N \oplus S_{i_0}$. Now consider the family $\mathcal{I}$ of all subsets J of I such that $N + (\sum_{j \in J} S_j)$ is direct. Since $\{i_0\} \in \mathcal{I}$, we find that $\mathcal{I} \ne \emptyset$. Proceeding exactly as above, we can find a subset J of I such that J is maximal for the property that $N + P = N \oplus P$ where $P = \sum_{j \in J} S_j$. If $N + P \ne M$, we can find, as above, a $j_0 \in I \setminus J$ such that $N + P' = N \oplus P'$ where $P' = \sum_{j \in J'} S_j$ with $J' = J \cup \{j_0\}$, contradicting the maximality of J in $\mathcal{I}$. Hence we get that $M = N \oplus P$. $\diamondsuit$

2.1.8 Corollary: *A sum or a direct sum of a family of semi–simple modules is semi–simple.*

2.1.9 Theorem: *For a semi–simple module M, the following are equivalent.*
1. *M is Noetherian.*
2. *M is finitely generated.*
3. *M is of finite length.*
4. *M is Artinian.*

Proof: $(1) \Rightarrow (2)$ is obvious.

$(2) \Rightarrow (3)$: Let $M = \bigoplus_{i \in I} S_i$ with S_i simple. Suppose M is generated by a finite subset $X = \{x_j\}$, $1 \le j \le n$. For each j write $x_j = \sum_{j(i) \in I} x_{j(i)}$ with $x_{j(i)} \in S_{j(i)}$ and all but finitely many $x_{j(i)}$ are zero. If $J =$

$\{j(i) \in I \mid x_{j(i)} \neq 0 \text{ for all } j, 1 \leq j \leq n\}$, then it is clear that J is finite and that $M = \oplus_{j \in J} S_j$ which means that M is a module of finite length (of length equal to $|J|$, the cardinality of J).

$(3) \Rightarrow (4)$ is obvious.

$(4) \Rightarrow (1)$ is also clear because an infinite direct sum of non–zero modules cannot be Artinian. Once it is a finite direct sum of simple (hence Noetherian) modules, it is Noetherian as well. ◊

2.1.10 Remark: Direct product of a family of semi–simple (*even simple*) modules need *not* be semi–simple.

For example, take an infinite dimensional vector space V over a division ring D and let $R = \text{End}_D(V)$. Considering V as a left module over R under evaluation, we see that V is simple. If B is a D–basis of V, then $R \approx \prod_{b \in B} V_b$ where $V_b = V$, for all $b \in B$, as R–modules. By (2.1.9) above, we find that R is *not* semi–simple as an R–module, since R is finitely generated over R but it is *neither* Artinian *nor* Noetherian (Prove this).

2.1.11 Proposition: *Suppose a module M has a submodule N which is a direct summand and that both N and M/N are semi–simple, then M is semi–simple.*

Proof: This is trivial because if $M = N \oplus P$ and both N and $P = M/N$ are semi–simple, obviously M is semi–simple. ◊

2.1.12 Remark: If a module M is such that it has a submodule N with both N and M/N are semi–simple, then M need *not* be semi–simple. (Since such an N *need not* be a direct summand of M.)

To see this, consider the finite cyclic group C_{p^2} of order p^2 where p is a prime number. If a is a generator of this group, then the subgroup C_p generated by a^p and the quotient group C_{p^2}/C_p are both of order p and hence semi–simple (in fact simple). But C_{p^2} is not semi–simple because if $C_{p^2} = C_p \oplus C_{p^2}/C_p$, it would mean that every element of C_{p^2} is of order at most p which is absurd. ∎

2.2   Isotypical Components

Given a ring R, we denote by $S = S_R$, a *complete set* of mutually non–isomorphic simple left R–modules, i.e., every simple left R–module is isomorphic to one and only one member of S. By (1.1.11) above, we note that S is simply the set of all *isomorphism classes* of the family of quotient modules R/I as I varies over the set of all maximal left ideals of R.

If N is a simple submodule of an R–module M, then $N \approx S$ for some $S \in S$. We denote by S_M the family of all simple modules $S \in S_R$ such that S is isomorphic to some simple submodule of M.

2.2.1 Isotypical module: Given a simple R–module S, we say that a semi–simple R–module M is *isotypical of type S* if M is a sum of a family of simple modules each of which is isomorphic to S.

2.2.2 Proposition: *Let $M = \sum_{i \in I} S_i$ with each S_i simple. If N is any simple submodule of M, then $N \approx S_i$ for some $i \in I$.*

Proof: Let N be a simple submodule of M. Since M is semi–simple, there is a submodule P of M such that $M = N \oplus P$. We have $M/P \approx N$ is simple. Let $\eta_P : M \to M/P$ be the natural homomorphism. We have $N = \eta_P(M) = \eta_P(\sum_{i \in I} S_i) = \sum_{i \in I} \eta_P(S_i) \neq (0)$ and so $\eta_P(S_j) \neq (0)$ for some $j \in I$. This gives that the restriction of η_P to S_j is a non–zero homomorphism between the simple modules S_j and N and hence it is an isomorphism by Schur's lemma (1.1.12).   ◊

2.2.3 Corollary: *Let M be a semi–simple module, isotypical of type $S \in S_M$. Then every simple submodule of M is isomorphic to S.*

2.2.4 Proposition: *Let M be a semi–simple module. For each $S \in S_M$, let M_S be the sum of all simple submodules of M each of which is isomorphic to S. Then*
1. *M_S is the largest submodule of M, isotypical of type S, called the isotypical component of type S and*
2. *$M = \bigoplus_{S \in S_M} M_S$, i.e., M is the direct sum of its distinct isotypical components.*

Proof: Let $M = \oplus_{i \in I} S_i$ with S_i simple. For each $S \in \mathcal{S}$, let $I_S = \{i \in I \mid S_i \approx S\}$.

Claim: We have $M_S = \oplus_{i \in I_S} S_i$.

It is obvious that $M_{I_S} = \oplus_{i \in I_S} S_i \subseteq M_S$. To see the reverse inclusion; take any simple submodule N of M such that $N \approx S$. Now look at the natural homomorphism $\eta_{I_S} : M \to M/M_{I_S} = \oplus_{j \notin I_S} S_j = M'_{I_S}$. If $\eta_{I_S}(N) \neq (0)$, then $N \approx \eta_{I_S}(N)$ which is a simple submodule of M'_{I_S} and hence by (2.2.2) above, N is isomorphic to S_j for some $j \notin I_S$ which contradicts the definition of the set I_S because $S \approx N \approx S_j$ but $j \notin I_S$. Hence $\eta_{I_S}(N) = (0)$, i.e., $N \subseteq M_{I_S}$ or $M_S = M_{I_S}$, as required.

Obviously M_S is the maximal isotypical submodule of type S and clearly we have $M = \oplus_{S \in \mathcal{S}_M} M_{I_S} = \oplus_{S \in \mathcal{S}_M} M_S$, as required.   ◊

2.2.5 Theorem: *Let M be semi–simple. Then every isotypical component of M is stable under all R–linear endomorphisms of M. Conversely, any submodule of M which is stable under all endomorphisms is a sum of suitable isotypical components of M. (In particular, an isotypical component is a smallest non–zero submodule which is stable under all endomorphisms.)*

Proof: Let f be an endomorphism of M. Let M_S be an isotypical component of M of type $S \in \mathcal{S}_M$. Take any simple submodule N of M_S. We have $N \approx S$. If $f(N) \neq (0)$, by Schur's lemma (1.1.12), $S \approx N \approx f(N)$ is simple and so $f(N) \subseteq M_S$, hence we get that $f(M_S) \subseteq M_S$, as required.

Suppose P is a submodule of M stable under all endomorphisms of M. It suffices to prove that whenever P contains a simple submodule N of M, then P contains all simple submodules of M each of which is isomorphic to N. Say Q is a submodule of M isomorphic to N by means of an isomorphism $\varphi : N \to Q$. Since M is semi–simple, we can write $M = N \oplus N'$ for some submodule N' of M. If $\mathrm{pr}_N : M \to M/N' = N$ is the projection onto N, then $f = \varphi \circ \mathrm{pr}_N$ is an endomorphism of M such that $f(M) = f(N) = \varphi(N) = Q$ and so $Q = f(N) \subseteq P$ since $N \subseteq P$ and P is stable under f.   ■

2.3 Endomorphism Rings

In this section, we shall determine the rings of endomorphisms of semi–simple modules in terms of known rings such as the endomorphism rings of vector spaces over division rings. The key is to know the endomorphism rings of modules which are semi–simple and isotypical. They are indeed the endomorphism rings of vector spaces, as we shall see soon.

2.3.1 Proposition: *Let M be a semi–simple module with its isotypical components $\{M_S, S \in \mathcal{S}_M\}$. Then the ring of endomorphisms of M is given by $\mathrm{End}_R(M) \approx \prod_{S \in \mathcal{S}_M} \mathrm{End}_R(M_S)$.*

Proof: This is quite easy. Let $f \in \mathrm{End}_R(M)$. For each $S \in \mathcal{S}_M$, let f_S be the restriction of f to M_S so that f_S is an endomorphism of M_S by (2.2.5) above. Look at the mapping $f \mapsto (f_S)_{S \in \mathcal{S}_M} \mapsto \sum_{S \in \mathcal{S}_M} f_S = f$, where in the middle term f_S is the restriction of f to M_S and in the last sum f_S stands for the extension of f_S to M with 0 on all of $M_T, S \neq T \in \mathcal{S}_M$. We note that the possible infinite sum $\sum_{S \in \mathcal{S}_M} f_S$ makes sense as a map. This is clearly an isomorphism of rings, as required. $\diamond$

The ring $\mathrm{End}_R(M)$ is known once we know the rings $\mathrm{End}_R(M_S)$ for the isotypical components $M_S, S \in \mathcal{S}_M$ of M. We may therefore assume that M is itself isotypical of some type S. We set the following notation.

Let $M = \oplus_{i \in I} S_i$ with $S \approx S_i$, $\forall\, i \in I$. Fix isomorphisms $\varphi_i : S \to S_i$ for all $i \in I$. We shall treat the φ_i as R–linear homomorphisms of S into M in a natural way, namely, $\varphi_i : S \to S_i \hookrightarrow M$.

Let $D = D_S = \mathrm{End}_R(S)$ which is a division ring by Schur's lemma (1.1.12) since S is simple. Let $V = V_S = \mathrm{Hom}_R(S, M)$, the space of all R–linear homomorphisms. This is an abelian group but not necessarily an R–module since R may not be commutative. However, V is a *right* vector space over D for the scalar multiplication given by $\varphi \colon S \to M$ and $\alpha \colon S \to S \Rightarrow \varphi \cdot \alpha = \varphi \circ \alpha \in V$.

2.3.2 Theorem: *The set $B = \{\varphi_i, i \in I\}$ is a D–basis of the (right)* *vector space V over D.*

Proof: (i) *B is linearly independent.*

Suppose we have a dependency relation $f = \sum_{r=1}^{n} \varphi_{i_r} \alpha_{i_r} \equiv 0$, i.e., $f(x) = \sum_{r=1}^{n} \varphi_{i_r}(\alpha_{i_r}(x)) = 0$ for all $x \in S$. Since $\varphi_{i_r}(\alpha_{i_r}(x)) \in S_{i_r}$, we get that $\varphi_{i_r}(\alpha_{i_r}(x)) = 0$ for all $x \in S$, $1 \leq r \leq n$. Since φ_{i_r} is an isomorphism onto S_{i_r}, we get that $\alpha_{i_r}(x) = 0$ for all $x \in S$, $1 \leq r \leq n$, i.e., $\alpha_{i_r} \equiv 0$, $1 \leq r \leq n$, as required.

(ii) *B spans V over D.*

Let $f \in V$. Since S is simple and f is R–linear, the map f is completely determined if we know its value on any non–zero element $x \in S$. Fix an $x \in S$, $x \neq 0$. Since $M = \oplus_{i \in I} S_i$, we have $f(x) = \sum_{r=1}^{n} x_r$ with $x_r \in S_{i_r}$, $i_r \in I$, $1 \leq r \leq n$. If we define $\alpha_r = \varphi_{i_r}^{-1} \circ \mathrm{pr}_{i_r} \circ f$, $1 \leq r \leq n$, then it follows that·we have $f = \sum_{r=1}^{n} \varphi_{i_r} \circ \alpha_r$, as required. (Note that the choice of the elements $i_r \in I$ in the proof depends on the element $x \in S$ but it does not matter as explained already.) $\Diamond$

2.3.3 Theorem: *Let M be isotypical of type S and let V and D be as above. Then there is an isomorphism of rings, $\mathrm{End}_R(M) \approx \mathrm{End}_D(V)$.*

Proof: The idea of the proof is rather simple. Define

$$\lambda : \mathrm{End}_R(M) \to \mathrm{End}_D(V), \ F \mapsto \lambda_F, \lambda_F(f) = F \circ f, \ \forall f \in V.$$

It is easy to check that λ is indeed a homomorphism of rings.

1. *λ is a monomorphism.*

We have $\mathrm{Ker} \ \lambda = \{F \mid F \circ f = 0, \ \forall f \in V\} = \{F \mid F \circ \varphi_i = 0, \ \forall i \in I\}$. Since $S_i = \varphi_i(S)$, we have $F(S_i) = F(\varphi_i(S)) = (F \circ \varphi_i)(S) = 0$ for all $i \in I$, hence $F \equiv 0$ for all $F \in \mathrm{Ker} \ \lambda$, as required.

2. *λ is an epimorphism.*

If $\Phi \in \mathrm{End}_D(V)$, then $\lambda_F = \Phi$ where $F = \sum_{i \in I} F_i \in \mathrm{End}_R(M)$ with $F_i = f_i \circ \varphi_i^{-1}$ and $f_i = \Phi(\varphi_i)$, etc. $\Diamond$

2.3.4 Corollary: *Let M be isotypical of type S and also a module of finite length n. Then the ring $End_R(M)$ is isomorphic to the matrix ring $M_n(D) = End_D(V)$ where $V = Hom_R(S, M)$ and $D = End_R(S)$.*

Given $f \in End_R(M)$, define $f_{ij} \in D$ by $f_{ij} = \varphi_i^{-1} \circ pr_i \circ f \circ \varphi_j$, $1 \le i, j \le n$. The mapping $f \mapsto (f_{ij})$ is an isomorphism we are looking for. ∎

2.4 Semi–simple Rings

2.4.1 Semi–simple rings: A ring R is called *semi-simple* if R is semi–simple as a left module over itself, i.e., every left ideal of R is a direct summand (2.1.4).

2.4.2 Proposition: *A ring R is semi–simple $\Leftrightarrow$ R is a direct sum of finitely many minimal left ideals.*

Proof: This is immediate by (2.1.9) above. However, the proof is so simple that it is worth repeating. If R is semi–simple, R is a direct sum of a family of simple submodules, i.e., minimal left ideals, say $R = \oplus_{i \in I} S_i$. Now write $1 = \sum_{r=1}^n x_r$, with $x_r \in S_{i_r}$ for suitable $i_r \in I$, $1 \le r \le n$. Thus we have $x = (\sum_{r=1}^n x x_r) \in \sum_{r=1}^n S_{i_r}$ for all $x \in R$ and hence $R \subseteq \sum_{r=1}^n S_{i_r} \subseteq R$, i.e., $R = \sum_{r=1}^n S_{i_r} = \oplus_{r=1}^n S_{i_r}$, as required. The converse is obvious. ◊

2.4.3 Corollary: *Let R be semi–simple with $R = \oplus_{r=1}^n S_r$, with S_r minimal left ideals for all r, $1 \le r \le n$. Then every simple R-module, in particular, every minimal left ideal of R is isomorphic to one of these S_r's. Consequently, there are only finitely many mutually non–isomorphic simple modules over R, i.e., S_R is finite.*

2.4.4 Proposition: 1. *Isotypical components of a semi–simple ring are minimal 2–sided ideals (and hence they are finite in number).*
2. *Every 2–sided ideal is a sum of suitable isotypical components (and hence there are only finitely many 2–sided ideals).*

Proof: By (2.2.5) above, it suffices to prove that an ideal I of R is 2–sided $\Leftrightarrow I$ is stable under all R–linear endomorphisms of R. But it is trivial to see that an R–linear endomorphism of R is simply *right multiplication* by an element of R. In fact, if $f\colon R \to R$ is R-linear, then $f(x) = f(x \cdot 1) = xf(1)$ for all $x \in R$, which means that f is the right multiplication by $f(1)$ and conversely. Thus saying that I is a 2–sided ideal is the same as saying that I is a submodule (i.e., a left ideal) which is stable for right multiplication by elements of R. ◊

2.4.5 Remark: *For any ring R, semi–simple or not, the ring of R–linear endomorphisms of R is naturally anti–isomorphic to the ring R, i.e., isomorphic to the ring R^0 opposite to R (1.1.1).*

To see this: note that for all $f, g \in \mathrm{End}_R(R)$, we have

$$(f \circ g)(x) = f(g(x)) = f(xg(1)) = (xg(1))f(1) = x(g(1)f(1), \ \forall \ x \in R$$

and hence the mapping $f \mapsto f(1)$ is an isomorphism of rings $\mathrm{End}_R(R)$ and R^0, as required.

2.4.6 Remark: A semi–simple ring is of finite length, i.e., both Artinian and Noetherian. However, a ring of finite length need not be semi–simple. For example, the rings $\mathbb{Z}/m^n\mathbb{Z}$ are of finite length for all m and $n \in \mathbb{N}$, but not semi–simple if $m, n \geq 2$ ((2.4.8)(2) below).

2.4.7 Theorem: *For a ring R, the following are equivalent.*
1. *Every module over R is semi–simple.*
2. *R is semi–simple.*
3. *R is Artinian and without radical, i.e., $J(R) = (0)$.*
4. *R is Artinian and without nilpotent left (/right/2–sided) ideals.*
5. *R is Artinian and without nil left (/right/2–sided) ideals.*

Proof: We shall prove (1) ⇔ (2) and (2) ⇒ (3) ⇒ (4) ⇒ (5) ⇒ (2).

The implication (1) ⇒ (2) is obvious.

(2) ⇒ (1): Let M be any R–module. Write M as a quotient of a free R–module, say $F = \oplus_{x \in X} R_x$ where X is a set of generators for M and $R_x = R, \ \forall \ x \in X$. Since R is semi–simple, so is the direct sum F and hence also its quotient M, as required.

(2) ⇒ (3): Since $R = \oplus_{r=1}^n S_r$ is a finite direct sum of minimal left ideals S_r, R is Artinian. We have only to show that $J(R) = (0)$. This

follows from the obvious fact that

$$(0) = \text{Ann}_R(R) = \bigcap_{r=1}^{n} \text{Ann}_R(S_r) \supseteq J(R).$$

$(3) \Rightarrow (4)$: This is obvious because $J(R)$ contains all nil ideals. Since a nil ideal in an Artinian ring is nilpotent, we get that $(4) \Rightarrow (5)$.

$(5) \Rightarrow (2)$: Since R is Artinian, we know that $J(R)$ is nilpotent and hence nil and so $J(R) = (0)$. Moreover, $J(R)$ is an intersection of finitely many maximal left ideals, say $(0) = J(R) = \cap_{r=1}^{n} M_r$. It is easy to see that the mapping, $\varphi: R \to \oplus_{r=1}^{n} R/M_r$, $x \mapsto (x + M_r)_{r=1}^{n}$ for all $x \in R$, is an R–linear monomorphism (in fact, an isomorphism by the *Chinese Remainder Theorem*) (Ex.(1.11.19) above). Thus R is isomorphic to a submodule of the finite direct sum of the simple modules, namely, $\oplus_{r=1}^{n} R/M_r$ and hence semi–simple.   ◇

2.4.8   Corollaries: 1. *If R is Artinian, then the quotient ring $R/J(R)$ (which is Artinian and without radical) is semi–simple.*
2. *The finite ring $\mathbb{Z}/n\mathbb{Z}$ (which is commutative Artinian) is semi–simple $\Leftrightarrow \mathbb{Z}/n\mathbb{Z}$ has no nilpotent elements $\Leftrightarrow n$ is square free.*

2.4.9   Theorem: *An Artinian ring is Noetherian* (i.e., *rings of finite length are nothing but Artinian rings*).

Proof: Let R be Artinian with its radical $J = J(R)$. Since R/J (being Artinian and without radical) is a semi–simple ring, it is of finite length and hence Noetherian. The result follows therefore if we prove that J is Noetherian as an R–module. To this end, we proceed as follows.

Since J is nilpotent, we have a finite descending chain of 2–sided ideals (starting with R and ending with (0)), namely,

$$R = J^0 \supseteq J \supseteq J^2 \supseteq \cdots \supseteq J^n = (0) \quad \text{for some } n \in \mathbb{N}.$$

Claim: The quotient modules $M_r = J^r/J^{r+1}$, $0 \leq r \leq n - 1$, are Noetherian.

This is a consequence of the following easy facts.
(a) *Each M_r is Artinian* (obvious) and (b) *each M_r is semi–simple.*

To see this, we note that each M_r is (obviously) annihilated by J and hence it can be treated as an R/J–module under the scalar multiplication $(a + J)x = ax$, $\forall\ a \in R$ and $\forall\ x \in M_r$. This is well–defined because $Jx = 0$, $\forall\ x \in M_r$. It is also clear that the two module structures on M_r over (R and R/J) are the same, i.e., each R–submodule is an R/J–submodule and vice–versa. But now R/J being semi–simple, any module over R/J is semi–simple. In particular, M_r is semi–simple as an R–module.

Thus M_r is Artinian and semi–simple and hence Noetherian by (2.1.9) above, as required.

Finally, we find that both $M_{n-1} = J^{n-1}/J^n = J^{n-1}$ and $M_{n-2} = J^{n-2}/J^{n-1}$ are Noetherian and hence J^{n-2} is Noetherian. Repeating this argument, we get by descending induction on n that for all r, $0 \le r \le n - 1$, J^r is Noetherian since both J^{r+1} and $M_r = J^r/J^{r+1}$ are Noetherian, as required. $\diamond$

2.4.10 Theorem (Artin–Molien–Wedderburn): *A ring R is semi–simple if and only if R is isomorphic to a finite direct product of matrix rings over suitable division rings.*

Proof: Suppose R is semi–simple. We know that $\mathcal{S}_R$ is finite, say $\mathcal{S}_R = \{S_i, 1 \le i \le n\}$. Write $R = \oplus_{i=1}^n R_i$ as the direct sum of its isotypical components R_i of type S_i. By (2.3.1) and (2.3.4) above, we get that $R^0 = \operatorname{End}_R(R) \approx \prod_{i=1}^n \operatorname{End}_R(R_i) = \prod_{i=1}^n M_{m_i}(D_i)$, where we have $D_i = \operatorname{End}_R(S_i)$, $m_i = \dim_{D_i}(V_i)$ and $V_i = \operatorname{Hom}_R(S_i, R_i)$ for $1 \le i \le n$.

Finally, we have $R = R^{00} = \prod_{i=1}^n (M_{m_i}(D_i))^0 = \prod_{i=1}^n M_{m_i}(D_i{}^0)$. But each $D_i{}^0$ is also a division ring. Thus R is a product of matrix rings over division rings. Conversely, a finite product of matrix rings over division rings is semi–simple since each is semi–simple (being Artinian and without radical). $\diamond$

Note: We have used the fact that for any ring R, the mapping $A \mapsto A^t$, the transpose of $A \in M_m(R)$, gives a natural isomorphism of the rings $(M_m(R))^0$ and $M_m(R^0)$ (Ex.(1.11.6) above).

2.4.11 Corollaries: 1. *A ring R is left semi–simple $\Leftrightarrow$ R is right semi–simple.*

2. *The centre of a semi–simple ring is a finite direct product of fields and is semi–simple itself.*

3. *A commutative ring is semi–simple $\Leftrightarrow$ it is a finite direct product of fields.*

Proof: (1) follows from the theorem since a matrix ring over a division ring is both left and right Artinian and without radical.

(2) We have

$$
\begin{aligned}
\text{Centre}(R) &= \text{Centre}(\prod_{i=1}^{n} M_{m_i}(D_i)) = \prod_{i=1}^{n} \text{Centre}(M_{m_i}(D_i)) \\
&= \prod_{i=1}^{n} \text{Centre}(D_i) = \prod_{i=1}^{n} K_i,
\end{aligned}
$$

which is a finite product of fields and obviously it is semi–simple.
(3) is obvious in view of (2).   ◇

Note: Unlike the case of semi–simple rings, left Artinian (resp. left Noetherian) rings need not be right Artinian (resp. right Noetherian). Counter examples to this effect are given in Exs.(1.11.37) and (1.11.38) above.

2.4.12 Theorem: *The isotypical components R_i of a non–zero semi–simple ring R are themselves rings with unity e_i where the e_i's are mutually orthogonal central idempotents forming a partition of unity and each R_i is a matrix ring over a division ring D_i.*

Proof: This is a reformulation of (2.4.10) above. In fact, if $R = \oplus_{i=1}^{n} R_i$ is the decomposition of R into its isotypical components R_i, then $R_i \approx R/R_i'$ which is a ring with unity where $R_i' = \oplus_{j \neq i} R_j$ (is a 2–sided ideal of R). Write $1 = \sum_{i=1}^{n} e_i$, $e_i \in R_i$ and observe that $e_i = (1 + R_i')$ is the unity of R_i. Further, we have $x_i x_j = x_j x_i = 0$ for all $x_i \in R_i$ and $x_j \in R_j$ $(1 \leq i \neq j \leq n)$ because both R_i and R_j are 2–sided ideals and hence $x_i x_j \in R_i \cap R_j = (0)$ for all $i \neq j$. In other words, we have $\text{Ann}_R(R_i) = R_i' = \oplus_{j \neq i} R_j$ for all $1 \leq i \leq n$, which means that the module structure of R_i as an R-module or as an R_i-module is the same. In particular, $e_i e_j = e_j e_i = 0$ for all $i \neq j$.

Now it follows that the e_i's are central since $x = \sum_{i=1}^{n} x e_i = \sum_{i=1}^{n} e_i x$ and hence $x e_i = x e_i^2 = e_i^2 x = e_i x$ for all $x \in R$. Finally, we have $R_i^0 \approx \text{End}_{R_i}(R_i) = \text{End}_R(R_i) = M_{m_i}(D_i)$ and hence $R_i \approx M_{m_i}(D_i^0)$, as required. ∎

2.5 Artinian Simple Rings

Recall (1.1.4) that a non–zero ring R is called a *simple ring* if R has no 2–sided ideals other than (0) and R. Some examples of simple rings are the matrix rings over division rings. These are Artinian too. However, there are simple rings which are not Artinian. For instance, let V be a countably infinite dimensional vector space over a division ring D and $S = \text{End}_D(V)$. Let I be the (unique) proper 2–sided ideal of S consisting of all endomorphisms of V of *finite rank*. Then the quotient ring $R = S/I$ is simple but not Artinian. (Prove this.)

In this section, we shall study Artinian simple rings leading to the well–known *structure theorem of Artin–Molien–Wedderburn* which says that these are nothing but the matrix rings over division rings. This is already implicit in the Artin–Molien–Wedderburn theorem on the structure of semi–simple rings ((2.4.10) and (2.4.12) above).

2.5.1 Theorem: *Let R be non–zero semi–simple ring. Then R is simple if and only if all simple modules over R are isomorphic to each other, i.e., $\mathcal{S}_R = \{S\}$ where S is a simple quotient of R.*

Proof: Since R is semi–simple, write $R = \oplus_{i=1}^{n} R_i$ as a direct sum of its isotypical components of type $S_i \in \mathcal{S}_R = \{S_1, \cdots S_n\}$. Since R is simple and each R_i is a 2–sided ideal of R, we must have $R_i = (0)$ or R. But $R_i \neq (0)$ unless $R = (0)$ and so $n = 1$, as required. Conversely, if all simple modules over R are isomorphic to each other, then $\mathcal{S}_R = \{S\}$ where S is a simple quotient of R and consequently, R is isotypical of type S, i.e., R is a minimal 2–sided ideal of R which means R is a simple ring, as required. ◇

2.5.2 Theorem (Artin–Molien–Wedderburn): *For a non–zero ring R, the following are equivalent.*

1. *R is Artinian and simple.*

2. *R is semi–simple and without proper two–sided ideals.*

3. *R is Artinian and isotypical.*

4. *R is a matrix ring over a division ring.*

Proof: (1) $\Rightarrow$ (2): Since R is simple. it is without two–sided ideals and hence its radical $J(R)$, being a two–sided ideal, must be zero. But then, R is Artinian and without radical and hence semi–simple by (2.4.7) above.

(2) $\Rightarrow$ (3): Since R is semi–simple, it is a direct sum of its isotypical components. But an isotypical component is a minimal two–sided ideal and so R must have a unique isotypical component, i.e., R is isotypical.

(3) $\Rightarrow$ (4): Obvious by (2.4.12) above.

(4) $\Rightarrow$ (1): Well–known. $\Diamond$

2.5.3 Theorem (Rigidity): *An Artinian simple ring R is isomorphic to a matrix ring $M_n(D)$ for a unique positive integer n and (up to an isomorphism) a uniquely determined division ring D.*

Proof: By the above theorem, $R \approx M_n(D)$ for some $n \in \mathbf{N}$ and a division ring D. We have only to prove that if $M_n(D) \approx M_{n'}(D')$, then $n = n'$ and $D \approx D'$. Comparing the lengths, we have $\ell_R(R) = n = n'$. On the other hand, if S is the unique simple R–module, unique up to isomorphism, then it suffices to show that $D^0 \approx \mathrm{End}_R(S)$ which would imply that $D^0 \approx (D')^0$, as required.

We may assume (up to an isomorphism) that S is the minimal left ideal of $R = M_n(D)$ consisting of all matrices whose columns are all zero except possibly the first. For each $\lambda \in D$, look at the *right multiplication* $r_{\lambda E_{11}}$, by the element λE_{11} of R, where E_{11} is the matrix having 1 at the $(1,1)^{\text{th}}$ place and zeros elsewhere. This is an R–linear map of R onto S and hence induces an R–linear endomorphism of S. Thus we have a mapping $\varphi\colon D \to \mathrm{End}_R(S)$, $\lambda \mapsto r_{\lambda E_{11}} = \varphi(\lambda)$ for all $\lambda \in D$. It is easy to check that this iꞏ an *anti–homomorphism* of rings, i.e., preserves addition but reverses multiplication. Since

$\varphi(\lambda)(E_{11}) = E_{11}\lambda E_{11} = \lambda E_{11}^2 = \lambda E_{11}$, it is clear that φ is a monomor-phism. Finallay, let $f \in \text{End}_R(S)$. Say, $f(E_{11}) = \sum_{i=1}^n \lambda_{i1}E_{i1}$. Since f is R–linear, we have $f(E_{11}) = f(E_{11}^2) = E_{11}f(E_{11}) = \lambda_{11}E_{11} = (r_{\lambda_{11}E_{11}})(E_{11})$. This shows that the R–linear endomorphism of S, namely, $g = f - r_{\lambda_{11}E_{11}}$ is such that $g(E_{11}) = 0$ and hence g is not an isomorphism of S but then by Schur's lemma we get $g \equiv 0$, i.e., $f = \varphi(\lambda_{11})$ and hence φ is an anti–isomorphism of rings. $\Diamond$

2.5.4 Corollary: *Let D be a division ring and M and N be two modules over $R = M_n(D)$ such that they have the same dimension (and finite) as vector spaces over D. Then M and N are isomorphic as R–modules.*

Proof: Let S be the unique simple R–module (unique up to isomor-phism). Since R is semi–simple, M and N are semi-simple over R and hence we can write $M = \underbrace{S \oplus \cdots \oplus S}_{r \text{ times}}$ and $N = \underbrace{S \oplus \cdots \oplus S}_{s \text{ times}}$ for some r and $s \in \mathbf{N}$. Now we have $\dim_D(M) = r \dim_D(S) = s \dim_D(S) = \dim_D(N)$, which implies that $r = s$ and hence $M \approx N$, as required. $\Diamond$

2.5.5 Corollary: *Given a semi–simple ring R, there exist unique positive integers n, m_r and division rings D_r $(1 \le r \le n)$, (unique up to isomorphism), such that $R \approx \prod_{r=1}^n M_{m_r}(D_r)$, a finite direct product of simple rings $M_{m_r}(D_r)$, called the simple factors of R. The simple factors are nothing but the isotypical components.*

2.5.6 Corollary: *The length of the ring $R = \prod_{r=1}^n M_{m_r}(D_r)$ is given by $\ell_R(R) = \sum_{r=1}^n \ell_R(M_{m_r}(D_r)) = \sum_{r=1}^n m_r$.*

2.5.7 Corollary: *Any two bases of a finitely generated free module over a semi–simple ring have the same cardinality (Ex.(1.11.40)).*

Proof: Let R be a semi–simple ring and F be a finitely generated free module over R. Write $F = \oplus_{i=1}^q S_i$ with S_i simple R–modules, $1 \le i \le q$. Hence F is of finite length over R and $\ell_R(F) = q$. On the other hand, if B is a basis of F having r elements, then we have $F = R \oplus \cdots \oplus R$ (r times) and hence $q = \ell_R(F) = r\ell_R(R)$. If B' is another basis of F having s elements, then we get $r\ell_R(R) = q = s\ell_R(R)$, i.e., $r = s$, as required. ∎

Though we do not need the results of the following <u>three</u> sections for the later developments in this book, it is nevertheless interesting to know the basic facts of the theory of Central Simple Algebras and Brauer Groups. We shall merely introduce the fundamental concepts and sketch the main ideas, leaving the details as exercises for the interested reader. Those who are interested only in the representations of groups may skip the rest of this chapter.

2.6   Dense Rings of Transformations

Let M be an R–module and $A = \text{End}_R(M)$ be the ring of R–linear endomorphisms of M. We know that M can be considered as a left A–module in a natural way under evaluations. Let $B = \text{End}_A(M)$. For $a \in R$, the left *homothecy* (i.e., scalar multiplication) by a, i.e., $\ell_a : x \mapsto ax$, $\forall\, x \in M$ is an element of $\text{End}_{\mathbb{Z}}(M)$ but need not be an element of A unless R is commutative. However, $\ell_a \in B$ because for all $\varphi \in A$ and $x \in M$, we have

$$\ell_a(\varphi(x)) = \ell_a(\varphi(x)) = a\varphi(x) = \varphi(ax) = \varphi(\ell_a(x)) = (\varphi\ell_a)(x).$$

This gives a map $\ell : R \to B, a \mapsto \ell_a$ which is obviously a homomorphism of rings.

It is interesting to note that the subring $\ell(R)$ of all homothecies in B is sufficiently big if M is semi–simple over R, in fact *dense* in B in the following sense.

2.6.1   Density Theorem (Jacobson): *Let M be a semi–simple R–module. Then R is dense in B in the sense that given $f \in B$ and finitely many elements $x_i \in M, 1 \leq i \leq n$, there exists an $a \in R$, depending on f and x_i's such that $f(x_i) = ax_i, 1 \leq i \leq n$. In other words, A-linear endomorphisms of M are like R–scalar multiplications on finite subsets of M.*

Since we will *not* be using this result, we omit the proof and refer the reader to [31] for details and also some corollaries of the theorem stated there. Just to indicate an idea of the proof, we shall only consider the *simplest case*, namely, $n = 1$. The general case can be reduced to this.

Since M is semi–simple, given $x \in M$, we can write $M = Rx \oplus N$ for some submodule N. Let pr_x be the projection of M onto Rx. Note that, for all $y \in M$, we have $\mathrm{pr}_x(y) = a_y x$, for some $a_y \in R$ and $a_x = 1$. Since $\mathrm{pr}_x \in A$ and $f \in B$, we have

$$f(x) = f(\mathrm{pr}_x(x)) = \mathrm{pr}_x(f(x)) = a_{f(x)}x, \quad \text{with } a_{f(x)} \in R,$$

as required. ∎

2.7 Central Simple Algebras

In this section, K denotes a *field* and an algebra A means an algebra over K which is also *finite dimensional* as a K–vector space. Recall (1.9.2) that K is a subring of the centre of A and furthermore A is an Artinian ring (being finite dimensional over K).

2.7.0 Simple algebra: We say that an algebra A is *simple* if the ring A is simple (1.1.4).

By (2.5.2) and (2.5.3) above, $A = M_n(D)$ for *unique* division algebra D (unique upto isomorphism) over K and *unique* positive integer n. This D is called the *division ring of A* and is denoted by D_A.

2.7.1 Central algebra: We say that a K–algebra A is *central* if the centre of A is K.

2.7.2 Central simple algebra: We say that a K–algebra A is *central simple* if A is both a central and a simple algebra.

Examples: 1. Any division ring D is a central simple algebra over $K = \mathrm{Centre}(D)$, called a *central division algebra*.
2. Any Artinian simple ring is a simple algebra over its centre.
3. Any algebra A is a central algebra over its centre.
4. The matrix ring $M_n(D)$ over a division ring D is a central simple algebra over the centre of D.

5. An algebra A is simple (resp. central) if and only if the *opposite* algebra A^0 (1.1.1) is so.

2.7.3 Theorem: *The base change of a central simple algebra by a simple algebra is simple.*

Proof: Let A be central simple and B be simple algebras over K. We have to show that $A \otimes_K B$ is simple. As noted above, $A = M_n(D_A)$ for unique *central division algebra* D_A over K. By Ex.(1.11.52) above, we have $M_n(D_A) \otimes B \approx M_n(D_A \otimes_K B)$ (as K–algebras). But then by Ex.(1.11.7) above, $M_n(C)$ is simple if $C = D_A \otimes_K B$ is simple. Thus we can assume that $A = D_A = D$ is a central division K–algebra.

Fix a K–basis $\{b_\lambda \in B | \lambda \in \Lambda\}$ for B. Recall that every element $z \in D \otimes_K B$ can be written as $z = \sum_{\lambda \in \Lambda} d_\lambda \otimes b_\lambda$ for unique $d_\lambda \in D$ with $d_\lambda = 0$ for all but finitely many λ's (by Ex.(1.11.46) above). Let $\Lambda_z = \{\lambda \in \Lambda | d_\lambda \neq 0\}$ which is a finite subset of Λ.

Let, if possible, I be a non–zero 2–sided ideal of $D \otimes_K B$. Choose a non–zero $z_0 \in I$ such that Λ_{z_0} is a *minimal element* in the set $\Lambda(I^\star)$ $= \{\Lambda_z | z \in I \text{ and } z \neq 0\}$. Writing $z_0 = \sum_{\lambda \in \Lambda_{z_0}} d_\lambda \otimes b_\lambda$, we can assume that $d_{\lambda_0} = 1$ for at least one $\lambda_0 \in \Lambda_{z_0}$ (by multiplying z_0 with a non–zero element of D if necessary). Since I is a 2–sided ideal of $D \otimes_K B$, we have $z_1 = (d \otimes 1)z_0 - z_0(d \otimes 1) = \sum_{\lambda \in \Lambda_{z_0}} (dd_\lambda - d_\lambda d) \otimes b_\lambda \in I$ for all $d \in D$. It is obvious that $\Lambda_{z_1} \subseteq \Lambda_{z_0}$ and $\lambda_0 \notin \Lambda_{z_1}$ since $d_{\lambda_0} = 1$. This contradicts the minimality of Λ_{z_0} in $\Lambda(I^\star)$ unless $z_1 = 0$. But then we get that $dd_\lambda = d_\lambda d$ for all $\lambda \in \Lambda$. This is true for all $d \in D$ and so we get that $d_\lambda \in K = \text{Centre}(D)$, i.e., $z_0 = \sum_{\lambda \in \Lambda_{z_0}} 1 \otimes d_\lambda b_\lambda \in (1 \otimes B)$. In other words, $z_0 \in I \cap (1 \otimes B)$ which is a 2–sided ideal of the subalgebra $1 \otimes B$. But $1 \otimes B$ is isomorphic to B by Ex.(2.9.8) below. This forces $z_0 = 0$, a contradiction, as required.   ◇

2.7.4 Remark: By Ex.(1.11.58) above, the hypothesis "central" in the theorem above cannot be dropped.

2.7.5 Commutant: By the *commutant* of a subset X of a ring A, we mean the subring $X' = \{a \in A \mid ax = xa, \ \forall \ x \in X\}$.

2.7.6 Bicommutant: By the *bicommutant* of a subset X of a ring A, we mean the commutant of the commutant of X, i.e., the subring $X'' = X'' = \{a \in A \mid ax' = x'a, \forall x' \in X'\}$. For example, the commutant of A and hence the bicommutant of (0) is the centre of A.

2.7.7 Theorem: *Let $X \subseteq A$ and $Y \subseteq B$ be subalgebras of K-algebras A and B. Then we have $(X \otimes_K Y)' = X' \otimes_K Y'$, where we have identified $X \otimes_K Y$ and $X' \otimes_K Y'$ with subalgebras of $A \otimes_K B$ (2.9.9).*

Proof: It is clear that $X' \otimes_K Y' \subseteq (X \otimes_K Y)'$. To see the reverse inclusion: fix a K-basis $\{b_\lambda \in B \mid \lambda \in \Lambda\}$ for B and write $z = \sum_{\lambda \in \Lambda} z_\lambda \otimes b_\lambda$ for any $z \in (X \otimes_K Y)'$. In particular, we have $(x \otimes 1)z = z(x \otimes 1)$, implying $x z_\lambda = z_\lambda x$ for all $x \in X$, i.e., $z_\lambda \in X'$, $\forall \lambda$ or $z \in X' \otimes_K B$. Similarly, we get that $z \in A \otimes_K Y'$ and therefore $z \in (X' \otimes_K B) \bigcap (A \otimes_K Y') = X' \otimes_K Y'$, as required. $\diamondsuit$

2.7.8 Corollaries: 1. *For all K-algebras A and B, we have*
$\mathrm{Centre}(A \otimes_K B) = \mathrm{Centre}(A) \otimes_K \mathrm{Centre}(B)$.
2. *Consequently, the tensor product of central simple algebras is central simple.*
3. *In particular, $A \otimes_K A^0$ is central simple if A is central simple.*

2.7.9 Proposition: *The dimension of a central simple algebra over its centre is a perfect square.*

Proof: Let A be a central simple K-algebra so that as noted above, $A = M_n(D_A)$ for unique central division K-algebra D_A and hence $\dim_K(A) = n^2 \dim_K(D_A)$. If K is *algebraically closed*, then D_A being finite dimensional division K-algebra, we get that $D_A = K$, proving the result. However, if K is *not* algebraically closed, base change A to an *algebraic closure* $\overline{K}$ of K (Ex.(2.9.16) below) to get the central simple $\overline{K}$-algebra $A \otimes_K \overline{K}$ and note that $\dim_K(A) = \dim_{\overline{K}}(A \otimes_K \overline{K})$ which is a perfect square, as required. $\diamondsuit$

2.7.10 Theorem: *For any central simple algebra A, we have a natural isomorphism of K-algebras, $\varphi \colon A \otimes_K A^0 \xrightarrow{\approx} \mathrm{End}_K(A)$, $a \otimes b \mapsto \ell_a r_b$, where ℓ_a is the K-endomorphism of A given by left multiplication by a and r_b is the right multiplication by b. In particular, the endomorphism ring of a central simple algebra is central simple.*

Proof: This is an immediate consequence of (1.10.18)(6) above, applied to the homomorphisms $\varphi_A\colon A \to \operatorname{End}_K(A)$, $a \mapsto \ell_a$ and $\varphi_{A^0}\colon A^0 \to \operatorname{End}_K(A)$, $b \mapsto r_b$, with the observations that

(i) $\ell_a \circ r_b = r_b \circ \ell_a$ for all $a, b \in A$,

(ii) the induced map φ is injective since $A \otimes_K A^0$ is simple and

(iii) $\dim_K(A \otimes_K A^0) = \dim_K(\operatorname{End}_K(A)) = (\dim_K(A))^2$, etc.   ◇

2.7.11 Corollary: *If A is a central simple algebra of dimension n, then $A \otimes_K A^0$ is central simple because $A \otimes_K A^0 \approx \operatorname{End}_K(A) \approx M_n(K)$.*

2.7.12 Theorem(Skolem–Noether): *Every K–automorphism of a central simple K–algebra A is an inner automorphism.*

More generally, if B is a simple K–algebra and $f, g\colon B \to A$ are monomorphisms of K–algebras, then there exists a unit $u \in A$ such that $g(x) = uf(x)u^{-1}$ for all $x \in B$.

Proof: Write $A = M_n(D_A)$ where D_A is the central division algebra of A. We proceed in two steps.

Step 1: Assume that $D_A = K$.

We know that K^n is a $M_n(K)$–module and hence we get two B–module structures on K^n by means of the K–monomorphisms f and g, namely, $bv = f(b)v$ and $bv = g(b)v$ for all $b \in B$ and $v \in K^n$. But $B = M_m(D)$ for some division K–algebra D and so by (2.5.4) above, these two B–module structures on K^n are isomorphic, i.e., there exists a K–automorphism, $u\colon K^n \to K^n$ such that $g(b)u(v) = u(f(b)v)$ for all $b \in B$ and $v \in K^n$, i.e., u is a unit in $M_n(K) = A$ such that $g = ufu^{-1}$, as required.

Step 2: A is arbitrary.

Let $F = f \otimes 1_{A^0}\colon B \otimes_K A^0 \to A \otimes_K A^0$ and $G = g \otimes 1_{A^0}\colon B \otimes_K A^0 \to A \otimes_K A^0$ which are K–monomorphisms of the simple algebra $B \otimes_R A^0$ into the central simple algebra $A \otimes_K A^0 = M_n(K)$ (2.7.11) where $n = \dim_K(A)$. But then by Step 1 above, there exists a unit $u \in A \otimes_K A^0$ such that $G = uFu^{-1}$. Hence we have

$$G(b \otimes a) = g(b) \otimes a = u(f(b) \otimes a)u^{-1}, \ \forall\, b \in B \text{ and } a \in A^0.$$

In particular, taking $b = 1$, we have $1 \otimes a = u(1 \otimes a)u^{-1}$ for all $a \in A^0$, i.e., $u \in (1 \otimes A^0)'$. On the other hand, by (2.7.7) above, we have

$$(1 \otimes A^0)' = \{1\}' \otimes (A^0)' = A \otimes K = A \otimes 1 \approx A.$$

Thus we get that u is a unit in A with the property that $g(b) = u f(b) u^{-1}$ for all $b \in B$, as required. $\Diamond$

2.7.13 Theorem: *Let B be a simple subalgebra of a central simple algebra A. Then we have the following.*
1. *The commutant B' of B is simple.*
2. *The bicommutant $B'' = B$ and*
3. $\dim_K(A) = \dim_K(B) \dim_K(B')$.

Proof: Let $m = \dim_K(B)$ so that $\operatorname{End}_K(B) \approx M_m(K)$ is central simple. The natural monomorphism $\eta_A: A \to A \otimes_K \operatorname{End}_K(B)$, $a \mapsto a \otimes 1$, gives, on restriction, a monomorphism $\eta_B: \eta_A|_B$ of B into the central simple algebra $A \otimes_K \operatorname{End}_K(B)$. On the other hand, left multiplication by elements of B gives a monomorphism $\ell: B \to \operatorname{End}_K(B)$, $b \mapsto \ell_b$, which in turn induces a monomorphism $\ell: B \to A \otimes_K \operatorname{End}_K(B)$, $b \mapsto 1 \otimes \ell_b$.

By Skolem–Noether, there exists a unit $u \in A \otimes_K \operatorname{End}_K(B)$ such that $\ell = u \eta_B u^{-1}$, hence $f(B) = u(B \otimes 1)u^{-1}$. Consequently, we get that $(\ell(B))' = u(B \otimes 1)'u^{-1} = u(B' \otimes_K \operatorname{End}_K(B))u^{-1}$. On the other hand, by Ex.(2.9.7), we have $(\ell(B))' = (1 \otimes \ell(B))' = \{1\}' \otimes (\ell(B))' = A \otimes_K B^0$. In particular, we have $A \otimes_K B^0 \approx B' \otimes_K \operatorname{End}_K(B)$ which implies B' is simple since ($A \otimes_K B^0$ simple by (2.7.3) above and hence) $B' \otimes_K \operatorname{End}_K(B)$ is simple. Equating dimensions, we get the required equality in (3) above. Finally, applying the same dimension formula (3) for the simple subalgebra B' in place of B, we get that $\dim_K(A) = \dim_K(B') \dim_K(B'')$ implying that $\dim_K(B) = \dim_K(B'')$. But $B \subseteq B''$ and so equality holds, as required. $\Diamond$

2.7.14 Corollary: *If B is a central simple subalgebra of a central simple algebra A, then B' is also central simple and the natural map $B \otimes_K B' \to A$, $x \otimes y \mapsto xy$, is an isomorphism.*

Obvious because the map is a monomorphism between simple algebras of the same dimension. $\Diamond$

2.7.15 Theorem: *Let $L \supseteq K$ be a (commutative) subfield of a central simple K-algebra A. Then the following are equivalent.*
1. *L is a maximal commutative subring of A.*
2. *$L = L'$.*
3. *$\dim_K(A) = (\dim_K(L))^2$.*

Proof: $(1) \Rightarrow (2)$: Since L is commutative, it is clear that $L \subseteq L'$. Moreover, for $x \in L'$, the subring $L[x]$ of A generated by L and x is commutative and contains L and so $L[x] = L$ by the maximality of L, i.e., $x \in L$ or $L = L'$.

$(2) \Rightarrow (3)$: Trivial by $(2.7.13)(3)$ above (since L, being a field, is a simple subalgebra of A).

$(3) \Rightarrow (1)$: Since L is commutative, $L \subseteq L'$. Hence by (3) and $(2.7.13)(3))$, we get that $L = L'$ (i.e., $(3) \Rightarrow (2)$). But any commutative subring of A containing L must be contained in $L' = L$, as required. $\diamond$

2.7.16 Corollary: *Let D be a central division K-algebra. Then D contains a maximal subfield $L \supseteq K$ with $\dim_K(D) = (\dim_K(L))^2$.*

Existence of a maximal commutative subring L of D is obvious for dimension reasons and any such L is a subfield serving the purpose.∎

2.8   The Brauer Group

As before, K stands for a field and only finite dimensional K-algebras are considered. For a central simple K-algebra A, we denote by D_A the division algebra of A so that $A = M_n(D_A)$, etc.

2.8.1 Brauer equivalence: Two central simple K-algebras A and B are said to be *Brauer equivalent* or simply *equivalent* and denoted $A \sim B$ if there exists an isomorphism of K-algebras between the matrix algebras $M_m(A)$ and $M_n(B)$ for some positive integers m and n. In symbols, we write $A \overset{K}{\sim} B$ or simply $A \sim B$ if $M_m(A) \approx M_n(B)$.

2.8.2 Remarks: It is easy to see the following.

1. $A \sim B$ if and only if $A \otimes_K M_m(K) \approx B \otimes_K M_n(K)$ as K–algebras for some $m, n \in \mathbb{N}$.

2. The relation "$\sim$" is an *equivalence relation* and obviously isomorphic algebras are equivalent whereas equivalent algebras need not be isomorphic as for example $K \sim M_n(K)$ for all $n \in \mathbb{N}$.

3. For all central simple K–algebras A, we have
(i) $A \sim M_n(A)$ for all $n \in \mathbb{N}$, (ii) $A \sim D_A$ and (iii) $K \sim A \otimes_K A^0$.

4. Combining with (2.5.3) above, we have $A \sim B \iff D_A \sim D_B \iff D_A \approx D_B$. Consequently, each equivalence class contains a *unique* (upto isomorphism) central simple division algebra.

5. If $A_i \sim B_i$, for $i = 1$ and 2, then $A_1 \otimes_K A_2 \sim B_1 \otimes_K B_2$, by Ex.(2.9.10) below.

Notation: The *set of equivalence classes* of central simple K–algebras under the Brauer equivalence $\sim$ is dented by $\mathrm{Br}(K)$. The equivalence class of a central simple K–algebra A is denoted by $[A]$.

Note that the set $\mathrm{Br}(K)$ is also same as the set of all *isomorphism classes* of central *division* algebras.

2.8.3 Brauer group: Given a field K, the set $\mathrm{Br}(K)$ of the equivalence classes of all finite dimensional central simple K–algebras is an abelian *group* under *tensor product* as (well–defined) *multiplication*, called the *Brauer group* of K.

We note that the identity element of $\mathrm{Br}(K)$ is $[K]$ and the inverse of $[A]$ is $[A^0]$ (since $[A][A^0] = [A \otimes_K A^0] = [M_n(K)] = [K]$).

Treating $\mathrm{Br}(K)$ *simply* as the set of isomorphism classes of central division algebras over K, it is *not* possible to define the group structure as above since the tensor product of central division algebras is central simple but *rarely* a division algebra.

2.8.4 Examples: 1. If K is *algebraically closed*, then $\mathrm{Br}(K) = 1$ since there are no finite dimensional division algebras over K other than K. In particular, $\mathrm{Br}(\mathbb{C}) = 1$.

2. $\mathrm{Br}(\mathbb{R}) = \{[\mathbb{R}], [\mathbb{H}]\} \approx \mathbb{Z}/2\mathbb{Z}$ (generated by $[\mathbb{H}]$) where $\mathbb{H}$ is the division algebra of *real quaternions* (Ex.(2.9.20) below).

3. $\mathrm{Br}(\mathbf{F}) = 1$ where $\mathbf{F}$ is any *finite field* (Ex.(2.9.21) below).

2.8.10 Brauer group of $\mathbb{Q}$: Determining the Brauer group of $\mathbb{Q}$ is a non–trivial task. It is a good project for an interested reader to study the proofs of the following Theorems 1 and 2 as it naturally provides exposure to some deep mathematics. We can describe the group $\mathrm{Br}(\mathbb{Q})$ as follows.

For a prime number p, let $\mathbb{Q}_p$ be the field of "*p–adic numbers*".

Theorem 1: $\mathrm{Br}(\mathbb{Q}_p) \approx \mathbb{Q}/\mathbb{Z}$ for all primes p.

Let us write $\mathbb{Q}_\infty = \mathbf{R}$ and identify $\mathrm{Br}(\mathbf{R}) = \mathbb{Z}/2\mathbb{Z}$ naturally with the subgroup of $\mathbb{Q}/\mathbb{Z}$ generated by $1/2$. Let P denote the set of *all positive primes together with the symbol* ∞ (the "infinite prime"). Consider the group $\bigoplus_{p \in P}\mathrm{Br}(\mathbb{Q}_p)$ and the natural map obtained by *addition* of coordinates as elements in $\mathbb{Q}/\mathbb{Z}$,

$$\Phi : \bigoplus_{p \in P} \mathrm{Br}(\mathbb{Q}_p) = \mathbb{Z}/2\mathbb{Z} \bigoplus_{\substack{p \in P \\ p \neq \infty}} \mathbb{Q}/\mathbb{Z} \longrightarrow \mathbb{Q}/\mathbb{Z}$$

which is obviously a surjective homomorphism. Now we have

Theorem 2: $\mathrm{Br}(\mathbb{Q}) = \mathrm{Ker}\, \Phi$, i.e., we have an exact sequence of groups:

$$0 \longrightarrow \mathrm{Br}(\mathbb{Q}) \longrightarrow \bigoplus_{p \in P} \mathrm{Br}(\mathbb{Q}_p) \overset{\Phi}{\longrightarrow} \mathbb{Q}/\mathbb{Z} \longrightarrow 0.$$

We shall end the section with a brief discussion on the "splitting fields" of central simple algebras. Given a field extension $L \supseteq K$, the base change to L gives a natural homomorphism of groups, $\mathrm{br}_{L/K}$: $\mathrm{Br}(K) \to \mathrm{Br}(L)$, $[A] \mapsto [L \otimes_K A]$.

2.8.11 Splitting fields: A field extension $L \supseteq K$ is called a *splitting field* for a central simple K–algebra A if the central simple L–algebra $L \otimes_K A$ is L–isomorphic to $M_n(L)$ for some $n \in \mathbf{N}$, i.e., $L \otimes_K A$ is Brauer–equivalent to L or equivalently, $[A] \in \mathrm{Ker}(\mathrm{br}_{L/K})$. We also say that L *splits* A or A *is split over* L.

For example, any algebraically closed field extension $L \supseteq K$ is a splitting field for all central simple K–algebras. More generally, if $L \supseteq K$

with $\text{Br}(L) = 1$, then L is a splitting field for all central simple K-algebras.

The proofs of the following are left for the reader.

2.8.12 Theorem: *Given a finite dimensional extension $L \supseteq K$ and a central simple K-algebra A, the following are equivalent.*
1. *L is a splitting field for A.*
2. *There exists a central simple K-algebra B such that $B \overset{K}{\sim} A$ and B contains L as a maximal commutative subring.*

In particular, any maximal subfield L of a central division K-algebra D is a splitting field for D as well as $M_m(D)$ for all $m \in \mathbb{N}$. ◇

2.8.13 Galois splitting fields: A splitting field $L \supseteq K$ for a central simple K-algebra A is called a *Galois splitting field* for A if L is a *Galois extension*, i.e., a *normal* and *separable* extension of K.

2.8.14 Theorem: Existence of Galois splitting fields.
1. *Every non-commutative central division K-algebra D contains a subfield $L \neq K$ which is separable over K.*
2. *Every central division K-algebra D contains a maximal subfield L which is separable over K, i.e., a separable splitting field for D.*
3. *Let $L(\supseteq K)$ be a separably algebraically closed field, i.e., L has no finite dimensional separable extensions other than L. Then L is a splitting field for all central simple K-algebras.*
4. *A central simple K-algebra A has a Galois splitting field L/K.* ■

2.9   Exercises

Unless otherwise stated explicitly, a ring means a ring with unity, commutative or not. When two or more ideals are referred to, all are supposed to be ideals of the same kind (left/right/2–sided). All modules and homomorphism of rings are assumed unitary.

1. Let R be a ring such that it is a direct sum of a family $\{I_\lambda \mid \lambda \in \Lambda\}$ of left ideals. Show that
   (i) Λ is finite, (say) $\Lambda = \{\lambda_1, \cdots, \lambda_n\}$,
   (ii) I_{λ_j} is generated by an idempotent e_j, $1 \leq j \leq n$ and
   (iii) $\{e_j, 1 \leq j \leq n\}$ is a set of pairwise mutually orthogonal idempotents forming a partition of unity.

   Conversely, show that any set of mutually pairwise orthogonal idempotents giving a partition of unity decomposes R into a direct sum of suitable left (resp. right) ideals.

2. Decompose the matrix ring $M_n(R)$ into a direct sum of n left (resp. right) ideals for any ring R. Describe the corresponding set of orthogonal idempotents.

3. Show that $R = M_n(\mathbf{Z})$ is not semi–simple for any $n \in \mathbf{N}$. (Hint: Minimal left ideals exist in a semi–simple ring.)

4. Let $R = \prod_{i=1}^r R_i$ be a semi–simple ring whose simple (i.e., isotypical) components are the R_i's. For an R–module M, let $M_i = R_i M$, $1 \leq i \leq r$. Show that M_i is either zero or else an isotypical component of M and that $M = \oplus_{i=1}^r M_i$.

5. Let I and J be *minimal* left ideals of a semi–simple ring R. Show that $IJ = J$ or (0) according as $I \approx J$ or not as R–modules.

6. Give an example of a ring A and a *commutative* subring C of A such that the bicommutant (2.8.5) $C'' \neq C$.

7. Let $\ell \colon A \to \mathrm{End}_{\mathbb{Z}}(A)$ be the map given by $a \mapsto \ell_a$ where $\ell_a(x) = ax$ is the left multiplication by a. Similarly let $r \colon A^0 \to \mathrm{End}_{\mathbb{Z}}(A)$, $a \mapsto r_a$, be the right multiplications. Prove the following.
   (i)   Both ℓ and r are monomorphisms.
   (ii) $\ell_a \circ r_b = r_b \circ \ell_a$ for all $a, b \in A$.
   (iii) $f \in \mathrm{End}_{\mathbb{Z}}(A)$ is A–linear if and only if $f = r_a$ for some $a \in A$.
   Consequently, (iv) $(\ell(A))' = r(A^0)$ and $(\ell(A))'' = \ell(A)$.

8. Let A and B be algebras over a commutative ring R. Show that the natural map $\eta_A\colon A \to A \otimes_R B$, $a \mapsto a \otimes 1$ is a *monomorphism* of R-algebras if B is a free R-module in which 1 can be extended to a basis (viz., when R is a field).

9. Let $f_i\colon A_i \to B_i$ be *monomorphisms* of R-algebras, $i = 1, 2$. Give examples to show that the tensor product $f_1 \otimes f_2\colon A_1 \otimes_R A_2 \to B_1 \otimes_R B_2$ need *not* be a monomorphism even if one of them is an isomorphism (1.10.7). Yet, show that $f_1 \otimes f_2$ is a monomorphism if R is a field.

10. Given a commutative ring R and positive integers m and n, show that the R-algebras $M_m(R) \otimes_R M_n(R) \approx M_{mn}(R) \approx M_m(M_n(R)) \approx M_n(M_m(R))$ are mutually isomorphic to each other. Furthermore, using the commutativity and associativity of tensor product, show that if $M_{m_i}(A_i) \approx M_{n_i}(B_i)$ for $i = 1$ and 2, then $M_k(A_1 \otimes_R A_2) \approx M_\ell(B_1 \otimes_R B_2)$ for some k and ℓ where the A_i and B_i are R-algebras.

11. Let A be a central K-algebra and $L \supseteq K$ be a field extension. Prove from first principles that $A \otimes_K L$ is a central algebra over L. (Hint: $\sum a_i \otimes b_i$ is central in $A \otimes_K L \iff a_i$ is central in A for all i.)

12. Let $f_i \in \operatorname{End}_K(V_i)$, where each V_i is a finite dimensional vector space over a field K, of dimension m_i, $i = 1, 2$. Show that
    (i) $\operatorname{trace}(f_1 \otimes f_2) = \operatorname{trace}(f_1) \operatorname{trace}(f_2)$ and
    (ii) $\det(f_1 \otimes f_2) = \det(f_1)^{m_2} \det(f_2)^{m_1}$.

13. Let $\mathbf{H} = \mathbf{H_R} = \mathbf{R}[i, j, k]$ be the division algebra of *real quaternions* (with $i^2 = j^2 = k^2 = -1$; $ij = k = -ji$; $jk = i = -kj$ and $ki = j = -ik$, etc.). Given $x = a + bi + cj + dk \in \mathbf{H}$, the element $\bar{x} = a - bi - cj - dk \in \mathbf{H}$ is called the *conjugate* of x. Show that the map $x \mapsto \bar{x}$ gives an isomorphism of $\mathbf{H}$ onto $\mathbf{H}^0$ as $\mathbf{R}$-algebras. Deduce that $[\mathbf{H}]$ is an element of order 2 in $\operatorname{Br}(\mathbf{R})$.

14. State and prove the "*Fundamental Theorem of Algebra*".

15. The following are equivalent for a field K.
    (i)   The fundamental theorem of algebra holds for K.
    (ii) Every irreducible polynomial in $K[X]$ is linear.
    (iii) There are no finite dimensional field extensions of K other than K itself.

    A field K is said to be *algebraically closed* if one (and hence all) of the above *equivalent* properties holds. Over such a field K, deduce that there are no finite dimensional division algebras other than K itself. Is the converse true?

16. Show that any field K has an *algebraic closure* $\overline{K}$ which is *unique* upto a K–isomorphism. (Recall that L is an algebraic closure of K means that (i) L is algebraically closed and (ii) L is *algebraic* (i.e., every element of L is a root of a non–constant polynomial) over K.)

17. Show that the field of all *algebraic numbers*, i.e., $\overline{Q}$, is the union of all finite dimensional extensions of Q in C.

18. Given a finite field F, show that there exists a *strictly ascending* infinite tower of finite dimensional extensions $F = F_0 \subset F_1 \subset \cdots \subset F_n \subset \cdots$ such that $\overline{F} = \cup_{n=0}^\infty F_n$.

19. Using the fact that C is algebraically closed, show that the only irreducible polynomials in $R[X]$ are linear or quadratics whose discriminants are negative. Deduce that the only finite dimensional field extensions of R are R and C and hence C is an algebraic closure of R.

20. Prove that (upto isomorphism) the only finite dimensional division R–algebras are R, C and H. (Hints: (i) If D is non–commutative, any maximal subfield L of D is C, (ii) $\dim_R(D) = 4$ by (2.8.16), (iii) The automorphism $z \mapsto \overline{z}$ of C is the restriction of an inner automorphism of D, say given by u, i.e., $\overline{z} = uzu^{-1}$ and so $z = u^2zu^{-2}$, $\forall\, z \in C$ $\Rightarrow u^2 \in C$, (iv) $u^2 \in R$, (v) $u^2 = -a < 0$, (vi) take $j = u/\sqrt{a}$ and $k = ij$ and (vii) $D = R[i, j, k] = H$.)

21. Prove that the "*norm map*" of a finite field over any of its subfields is surjective.

22. Using the method of proof of the previous two exercises, prove the well–known theorem that "*Every finite division ring is commutative, i.e., a field*".

23. **Schur's Commutation Theorem:** This gives a complete characterisation of all semi–simple (resp. simple) subalgebras of the matrix algebra $A = M_n(K)$ over an *algebraically closed field* K.

    Theorem: *Given a semi–simple subalgebra B of A, there exist positive integers r, ℓ_i, m_i, $1 \le i \le r$, such that the following are true.*
    (i) $n = \sum_{i=1}^r \ell_i m_i$.
    (ii) *The mapping $\phi\colon \prod_{i=1}^r M_{m_i}(K) \to M_n(K)$, sending*

$$(P_1, \cdots, P_r) \mapsto \mathrm{diag}(P_1^{\ell_1}, \cdots, P_i^{\ell_i}, \cdots, P_r^{\ell_r})$$
$$= \mathrm{diag}(\underbrace{P_1, \cdots, P_1}_{\ell_1 \text{ times}}, \cdots, \underbrace{P_i, \cdots, P_i}_{\ell_i \text{ times}}, \cdots, \underbrace{P_r, \cdots, P_r}_{\ell_r \text{ times}})$$

is an isomorphism of K-algebras <u>onto</u> *B.*

Let us write $B = \mathrm{Image}(\phi) = \prod_{i=1}^{r} \Delta_{\ell_i}(M_{m_i}(K))$.

(iii) *The commutant of B is given by $B' = \prod_{i=1}^{r} \Delta_{m_i}(M_{\ell_i}(K))$, interchanging the roles of ℓ_i and m_i for all i.*

(iv) *B is simple if and only if $r = 1$. In particular, the commutant of a simple (resp. semi-simple) subalgebra of A is one such.*

(v) *The set of isomorphism classes of all simple subalgebras of A is bijective with the set of all divisors of n.*

Which parts of this theorem fail if K is *not* algebraically closed? (Hint: As a B-module, decompose the A-module $V = K^n$ into the sum of its isotypical components, say $V = \oplus_{i=1}^{r} V_i^{\oplus \ell_i}$ where the V_i's are the mutually non-isomorphic B-simple submodules with $\dim_K(V_i) = m_i$, $1 \le i \le r$.) ∎

2.10   True/False Statements

Determine which of the following statements are true (T) or false (F) or partially true (PT). Justify your answers by giving a proof if (T) or providing a counter-example if (F)/(PT) or supplying the additional hypothesis needed to make it (T) (along with a proof) if (PT), as the case may be.

1. A vector space is semi-simple.

2. The matrix ring $M_n(D)$ over a division ring D is both a simple ring and a semi-simple module over itself.

3. Quotient of a semi-simple module is isomorphic to a submodule.

4. A local ring is semi-simple only if it is a division ring.

5. A finite ring is a semi-simple ring.

6. If a module M has a submodule N such that N and M/N are simple, then M is semi-simple.

7. Direct product of semi-simple modules is semi-simple.

8. Direct product of simple modules is semi-simple.

9. A simple ring is isomorphic to a matrix ring over a division ring.

10. An Artinian simple ring is a matrix ring over a division ring.

11. A semi–simple ring is Artinian without 2–sided ideals.

12. A semi–simple ring is Artinian without nilpotent ideals.

13. An Artinian simple ring is without left/right ideals.

14. A left Artinian ring is right Artinian.

15. A left Artinian simple ring is right Artinian.

16. A semi–simple ring without nilpotent ideals is simple.

17. The Jacobson radical of a semi–simple ring is (0).

18. A semi–simple ring is a product of finitely many simple rings.

19. Endomorphism ring of a semi–simple module is semi–simple.

20. Tensor product (over $\mathbf{Z}$) of division rings is a simple ring.

21. Tensor product of semi–simple rings is semi–simple.

22. Tensor product of central K–algebras is central.

23. Tensor product of central simple algebras is simple.

24. The commutant of a simple subalgebra is simple.

25. The bicommutant of a simple subalgebra is itself.

26. A semi–simple subalgebras of $A = M_p(\mathbf{C})$ is $\mathbf{C}$ or A for all primes p.

27. Every automorphism of a division algebra is inner.

28. Every automorphism of a central division algebra is inner.

29. Every automorphism of a central simple algebra is inner.

30. A central simple algebra over K is $M_n(K)$ for some $n \in \mathbf{N}$.

31. Any automorphism of $\mathbf{R}$ is the identity.

32. Any automorphism of $\mathbf{C}$ is the identity.

33. Any $\mathbf{R}$–automorphism of $\mathbf{C}$ is an inner automorphism.

34. A field whose Brauer group is trivial is finite or algebraically closed.

35. Fields with isomorphic Brauer groups are mutually isomorphic. ∎

Part II

REPRESENTATIONS OF FINITE GROUPS

Chapter 3

Representations of Finite Groups

In this chapter, we shall study the central theme of this book, namely, *representation of a finite group G* over a field K. We establish some of the basic properties using the theory of semi–simple modules over the so called *"group algebra"* of G over K. (We shall illustrate these results for several important special cases of groups such as S_n(Chapter 5), A_n(Chapter 6), B_n(Chapter 7) and D_n(Chapter 8) below.)

As an application of these results, we deduce some properties of the group G itself. We shall prove, for instance, a theorem of Burnside which says that a finite group of order $p^a q^b$ with p and q primes, is *solvable* (3.9.6). To do this, we work over $K = \mathbb{C}$, the field of complex numbers. Two important facts that play crucial role in the proof are that $\mathbb{C}$ is *algebraically closed* and of characteristic 0. To gain a little more generality, we will work over an arbitrary algebraically closed field K, often of characteristic 0. However, in what follows, the field $\mathbb{C}$ may be kept in place of K, if necessary. (Cf. [10], [23], [31], [42], [46], etc.)

3.1   The Group Algebra

Recall (1.9.1) that by an algebra A over a commutative ring R, we mean a ring A containing R as a central subring. We assume also that the unity of R is the same as that of A. In what follows, we will be *primarily* concerned with finite groups and algebras over fields. All groups will be multiplicatively written even if they are abelian.

3.1.1 Group algebra: Let R be a commutative ring (with unity) and G a group. By the *group algebra* of G over R, we mean the R-algebra $R[G]$ where $R[G]$ is a *free* module over R with the *set* G as an R-basis and the multiplication being given by

$$\left(\sum_{x \in G} a_x x\right) \cdot \left(\sum_{y \in G} b_y y\right) = \sum_{z \in G}\left(\sum_{xy=z} a_x b_y\right) z$$

with a_x, $b_y \in R$ and all but finitely many a_x's (resp. b_y's) are 0.

In case G is a finite group, say $G = \{x_1, \cdots, x_n\}$, we have
(i) $R[G] = \{\sum_{i=1}^{n} a_i x_i \mid a_i \in R\}$ and
(ii) $(\sum_{i=1}^{n} a_i x_i) \cdot (\sum_{j=1}^{n} b_j x_j) = \sum_{k=1}^{n}(\sum_{x_i x_j = x_k} a_i b_j) x_k$.

Note: In case $R = \mathbb{Z}$, the group algebra $\mathbb{Z}[G]$ is sometimes called the *group ring* of G.

3.1.2 Remark: If e is the identity element of G, then e is the unity of $R[G]$ and the mapping $a \mapsto a \cdot e = ae$ of R into $R[G]$ gives a monomorphism of rings with $1 \mapsto e$ and the image which is isomorphic to R is a central subring of $R[G]$. We identify R with this central subring of $R[G]$ and treat $R[G]$ as an algebra over R. It is obvious that $R[G]$ is commutative if and only if G is abelian.

Hereafter, we shall denote the identity element of a multiplicatively written group (even if the group is abelian) by 1.

3.1.3 Remark: For an R-algebra A if $\mathcal{U}(A)$ denotes the *group of units* in A, then we see that G is a subgroup of $\mathcal{U}(R[G])$ and the sets $\text{Hom}_{\text{groups}}(G, \mathcal{U}(A))$ and $\text{Hom}_{R-\text{alg}}(R[G], A)$ are naturally bijective with each other.

3.1.4 Examples: 1. If $G = \{X^m \mid m \in \mathbb{Z}\}$ is an infinite cyclic group generated by X, then $R[G] = R[X, X^{-1}]$ is the ring of *Laurent polynomials* in the variable X over R (1.1.2)(6).
2. If G is a finite cyclic group of order n generated by x, then $R[G] = R[x] = R[X]/(X^n - 1)$. An R-basis of $R[G]$ is the set $G = \{x, x^2, \cdots, x^{n-1}, x^n = 1\}$ where x is the residue of X modulo the ideal $(X^n - 1)$.

3.1.5 Remark: If G contains a non–trivial element of finite order (i.e., a non–trivial *torsion element*), in particular if G is a finite group of order ≥ 2, then $R[G]$ is *not* an integral domain even if R is a field.

For, let $x \in G$ be an element of finite order $m \geq 2$. Now $z_1 z_2 = 0$ where $z_1 = x - 1 \neq 0$ and $z_2 = 1 + x + \cdots + x^{m-1} \neq 0$ in $R[G]$, as required.

Note: In order that $R[G]$ be an *integral domain*, it is necessary that R is an integral domain and G is *torsion free*. However, these necessary conditions are, in general, *not* sufficient. (Give an example.)

3.1.6 Theorem: *Let G be a finite group of order n with its conjugacy classes $C_i, 1 \leq i \leq h$. Let $z_i = \sum_{x \in C_i} x$. Then $Z = \{z_i, 1 \leq i \leq h\}$ is an R–basis of $C = \mathrm{Centre}(R[G])$.*

Proof: It is clear that Z is linearly independent over R. Since $y C_i y^{-1} = C_i$ for all i and $y \in G$, we have $y z_i y^{-1} = z_i$, i.e., $Z \subseteq C$. We have only to prove that Z spans C over R. Let $z \in C$ and write $z = \sum_{x \in G} a_x x$ with $a_x \in R$. Fix an $x \in G$ such that $a_x \neq 0$. Take any $y \in C_x$, the conjugacy class through x, say $y = g x g^{-1}$ for some $g \in G$. Now we have $z = g z g^{-1} = \sum_{u \in G} g a_u u g^{-1} = a_x g x g^{-1} + \sum_{u \neq x} g a_u u g^{-1} = = a_x y + \cdots = a_x x + \cdots$, which implies that $a_y = a_x$ for all $y \in C_x$. Writing $a_i = a_x$ for $x \in C_i$, $1 \leq i \leq h$, we see that $z = \sum_{i=1}^{h} a_i z_i$. $\Diamond$

In case G is not finite, the above result takes the following form.

Theorem: *Let G be a group with the set of all its conjugacy classes $\{C_i \mid i \in I\}$. Let $J = \{j \in I \mid C_j$ is finite $\}$. (For instance, $J = I$ if G is finite or abelian). For each $j \in J$, let $z_j = \sum_{x \in C_j} x$ so that $z_j \in R[G]$. Then the set $Z = \{z_j \mid j \in J\}$ is an R–basis of $C = \mathrm{Centre}(R[G])$.*

Proof: We have only to prove that Z spans C. Let $z \in C$. As an element of $R[G]$, we can uniquely write $z = \sum_{x \in G} a_x x$ with $a_x \in R$ and all but finitely many a_x's are zero. The set $\mathrm{supp}(z) = \{x \in G \mid a_x \neq 0\}$ (which is a finite subset of G) is called the *support* of z.

Claim: For each x in $\mathrm{supp}(z)$, we have the following.

1. The conjugacy class C_x through x in G is finite,

2. $C_x \subseteq \mathrm{supp}(z)$ and

3. $a_x = a_y$ for all $y \in C_x$.

For, otherwise, for some x' in $\mathrm{supp}(z)$, the conjugacy class $C_{x'}$ of x' is infinite. Pick an element $y' \in C_{x'}-(\mathrm{supp}(z))$. We have $y' = tx't^{-1}$ for some $t \in G$. Since $z \in C$, we have $zg = gz$, $\forall\ g \in G$ hence, in particular, $zt = tz$. Now we have $z = tzt^{-1} = a_{x'}y' + \cdots$ which implies by uniqueness of expression for z that $a_{y'} = a_{x'} \neq 0$ and so $y' = hx'h^{-1}$ in $\mathrm{supp}(z)$. But this contradicts the choice of y'. This proves (1). Since we have shown that $y \in C_x \Rightarrow a_y = a_x$ and this is true for all x in $\mathrm{supp}(z)$, we get (2) and (3) as well.

Lastly, decompose $\mathrm{supp}(z) = \cup_{r=1}^{m}C_{j_r}$ into the union of the conjugacy classes for some $j_r \in J, 1 \leq r \leq m$. Let $a_r = a_x, x \in C_{j_r}$. Now we have $z = \sum_{r=1}^{m} a_r z_{j_r}$, as required.   ◇

Henceforward, G denotes a finite group of order n and K a field of arbitrary characteristic (p, i.e., p = 0 or else a prime number p).

3.1.7 Theorem (Maschke): *Let G be a finite group of order n and K a field. Then the group algebra $K[G]$ is semi–simple if and only if $n \cdot 1 \neq 0$ in K, i.e., the characteristic of K, if non–zero, does not divide n.*

Proof: Suppose $K[G]$ is semi–simple. Then, by (2.4.7) above, we get that there are no non–zero nilpotent ideals in $K[G]$. Suppose Char $K = p \neq 0$. Let, if possible, $n \cdot 1 = 0$ in K, i.e., $p \mid n$. Now look at the non–zero element $z = \sum_{x \in G} x$ which is a central element of $K[G]$. We have $xz = \sum_{y \in G} xy = \sum_{u \in G} u = z$ for all $x \in G$ (since $xG = G$). Hence we have $z^2 = (\sum_{x \in G} x)^2 = \sum_{x \in G} xz = nz = 0$. This implies that the non–zero principal ideal (z) in $K[G]$ is of square (0) since z is central. This is a contradiction and hence $n \neq 0$ in K, as required.

Conversely, suppose $n \neq 0$ in K. We write $(1/n)$ for the inverse of $n \cdot 1$ in K. We shall prove that any left ideal I of $K[G]$ is a direct summand. We may assume that $I \neq (0)$. Since $K[G]$ is a vector space over K, I is a vector subspace of $K[G]$ and hence there is a projection $f : K[G] \to I$ of vector spaces, i.e., f is only K–linear. For $z \in K[G]$, let $\ell_z : K[G] \to K[G]$ be the left multiplication by z. Now we look at the *average* $\widetilde{f}$ of f over the group G (defined below) and find that $\widetilde{f}$ is indeed a $K[G]$–linear projection we are looking for.

Claim: $\widetilde{f} \overset{\text{def}}{=} (1/n)\sum_{x \in G}(\ell_{x^{-1}} \circ f \circ \ell_x)$ is a $K[G]$–linear projection of $K[G]$ onto I (hence I is a direct summand of $K[G]$).

Let $z \in I$ and $y \in G$. We have only to show that $\widetilde{f}(z) = z$, $\widetilde{f}(yz) = y\,\widetilde{f}(z)$ and Image $\widetilde{f} \subseteq I$. We have

$$
\begin{aligned}
\widetilde{f}(z) &= \frac{1}{n}\sum_{x \in G}(\ell_{x^{-1}} \circ f \circ \ell_x)(z) = \frac{1}{n}\sum_{x \in G}(x^{-1}f(xz)) \\
&= \frac{1}{n}\sum_{x \in G}(x^{-1}xz) \text{ (since } xz \in I, f(v) = v, \; \forall\, v \in I) \\
&= \frac{1}{n}\sum_{x \in G} z = \frac{1}{n}nz = z.
\end{aligned}
$$

Secondly, we have

$$
\begin{aligned}
\widetilde{f}(yz) &= \frac{1}{n}\sum_{x \in G}(\ell_{x^{-1}} \circ f \circ \ell_x)(yz) = \frac{1}{n}\sum_{x \in G}x^{-1}f((xy)z) \\
&= \frac{1}{n}\sum_{x \in G}(yy^{-1})x^{-1}f((xy)z) = \frac{1}{n}\sum_{x \in G}(y(xy)^{-1}f((xy)z) \\
&= y\frac{1}{n}\sum_{xy=x' \in G}x'^{-1}f(x'z) = y\widetilde{f}(z).
\end{aligned}
$$

Lastly, for $v \in K[G]$ and $x \in G$, we have $f(xv) \in I$ and so

$$
\widetilde{f}(v) = \frac{1}{n}\sum_{x \in G}(\ell_{x^{-1}} \circ f \circ \ell_x)(v) = \frac{1}{n}\sum_{x \in G}x^{-1}f(xv) \in I. \qquad \Diamond
$$

3.1.8 Corollary: $K[G]$ *is semi–simple if and only if $K[H]$ is semi–simple for every subgroup H of G.*

3.1.9 Remark: $K[G]$ is a simple ring if and only if $G = \{1\}$ because $z = \sum_{x \in G}x$ is a central element and the 2–sided principal ideal $(z) = Kz \neq (0) \Rightarrow K[G] = Kz$ is of dimension 1 over K. ∎

3.2   Simple Modules over $K[G]$

Recall that G denotes a finite group of order n and K a field of characteristic 'p'. *We assume from now on that $K[G]$ is* semi–simple.

By (2.4.10), (2.4.12) and (2.5.3) above, write $K[G] = \prod_{i=1}^{r} M_{m_i}(D_i)$ for *uniquely* determined positive integers r, m_i and division rings D_i, $1 \leq i \leq r$. We also note that the integer r, which is the number of simple factors of $K[G]$, is the number of mutually non–isomorphic simple modules S_i over $K[G]$, i.e., $\mathcal{S}_{K[G]} = \{S_i \mid 1 \leq i \leq r\}$ (a complete set of mutually non–isomorphic simple $K[G]$–modules) and $D_i = (\text{End}_{K[G]}(S_i))^0$, etc.

Since $K[G]$ is a K–algebra, each of its simple factors $M_{m_i}(D_i)$ is also a K–algebra which simply means that each of the division rings D_i is a K–algebra. Since $K[G]$ is of finite dimension n over K, it follows that each D_i is finite dimensional over K and we have

$$n = \dim_K(K[G]) = \sum_{i=1}^{r} m_i^2 \dim_K(D_i) = \sum_{i=1}^{r} m_i^2 d_i$$

where $d_i = \dim_K(D_i), 1 \leq i \leq r$.

Suppose $K_i = \text{Centre}(D_i)$, $1 \leq i \leq r$. Then each K_i is a finite dimensional field extension of K and we have
(i) $d_i = \dim_K(D_i) = \dim_K(K_i) \dim_{K_i}(D_i)$ and hence
(ii) $n = \sum_{i=1}^{r} m_i^2 d_i = \sum_{i=1}^{r} m_i^2 (\dim_K K_i)(\dim_{K_i} D_i)$.

Since $\text{Centre}(K[G]) = \prod_{i=1}^{r} K_i$, we get the following.

3.2.1 Theorem: *Let h be the number of conjugacy classes of G, then we have $h = \dim_K(\text{Centre}(K[G])) = \sum_{i=1}^{r} \dim_K K_i$. Consequently, $r \leq h$ and equality holds $\Longleftrightarrow$ $K = K_i$ for all i, $1 \leq i \leq r$.*   ◇

In case K is *algebraically closed*, we know that there are *no* finite dimensional division algebras over K other than K itself (Ex.(2.9.15) above). We have then $K = K_i = D_i$ and hence we have the following *summary of facts* needed in the sequel.

3.2.2 Theorem: *Let G be a finite group of order n. Let K be an algebraically closed field such that $K[G]$ is semi–simple (for example,*

$K = \mathbb{C}$). *Let h be the number of conjugacy classes of G. Then there are exactly h mutually non–isomorphic minimal left ideals S_i in $K[G]$ and any simple $K[G]$–module is isomorphic to one of them. Writing $K[G] = \prod_{i=1}^{h} M_{m_i}(K)$, we get that each of the simple factors $M_{m_i}(K)$ is isotypical of type S_i and S_i can be taken to be the set of all matrices in $M_{m_i}(K)$ all of whose columns but the first are zero. Finally we have $n = \sum_{i=1}^{h} m_i{}^2$ where $m_i = \dim_K(S_i)$, $1 \le i \le h$.* ◇

3.2.3 Corollary: *With the same assumptions as above, i.e., the semi–simplicity of $K[G]$ with K being algebraically closed, the following are equivalent.*
1. *G is abelian.*
2. *Every simple $K[G]$–module is of dimension 1 over K.*
3. *$K[G]$ is a finite direct product of fields each containing K.*
4. *$K[G]$ is commutative.*

3.2.4 Corollary: *Let S be a simple $K[G]$–module. Let A be an abelian subgroup of G. Then $\dim_K(S) \le [G : A]$. (See also (3.8.9), (3.10.7) and Ex.(4.8.10) below.)*

Proof: Let T be a simple $K[A]$–submodule of S so that we have $\dim_K(T) = 1$. Since S is $K[G]$–simple, we get that $K[G]T = S$. Furthermore, for $x \in G$ and $a \in A$, we have $xaT \subseteq xT$ and so $xAT = xT$. Hence it follows that $S = K[G]T = \sum_{x \in G/A} xT$. But this is a sum of $[G : A]$ (and hence a direct sum of at the most $[G : A]$) one dimensional subspaces xT of S, as required. ◇

3.2.5 Remark: Most of what is stated in the above theorem and its corollaries is *not* true if K is *not* algebraically closed. For example, look at the case of $K = \mathbb{Q}$ and $G = \mathbb{Z}/p\mathbb{Z}$, the cyclic group of order p, where $p \ne 2$ is a fixed (odd) prime number. We have

$$
\begin{aligned}
\mathbb{Q}[\mathbb{Z}/p\mathbb{Z}] &= \mathbb{Q}[X]/(X^p - 1) \\
&\approx \big(\mathbb{Q}[X]/(X - 1) \big) \times \big(\mathbb{Q}[X]/(1 + X + \cdots + X^{p-1}) \big) \\
&\approx \mathbb{Q} \times \mathbb{Q}[\zeta_p] = S_1 \oplus S_2, \text{ say,}
\end{aligned}
$$

which is a direct sum of simple $\mathbb{Q}[\mathbb{Z}/p\mathbb{Z}]$–modules, where $\zeta_p = e^{2\pi i/p}$ is a *primitive* complex p^{th}–root of unity. Thus $r = 2$ whereas we know that $p = h$, i.e., $r < h$. On the other hand, we have $\dim_{\mathbb{Q}}(S_2) = m_2 = p - 1 \geq 2$ and lastly, $m_1 = 1$ and $m_2 = p - 1 \Rightarrow p \neq m_1{}^2 + m_2{}^2$, etc.

However, we note that the statements (1) and (4) of (3.2.3) above, are equivalent to the statement that
3′. $K[G]$ *is a finite direct product of fields each isomorphic to* K (*without assuming that* K *is algebraically closed*).

Cyclic modules over $K[G]$

Recall that a module generated by one element is called a *cyclic module* or a *monogenic module* and cyclic modules are nothing but quotients of $K[G]$. Simple modules are an important class of cyclic modules, being generated by any one of their non–zero elements. We shall conclude this section by giving a set of some sufficient conditions so as to describe a K–basis for a cyclic module over $K[G]$. The statement as well as its proof appears a little technical but useful. Since this is not required in the rest of this chapter, the reader may skip this part until its use in (5.7.13) below.

For an element $z = \sum_{x \in G} a_x x \in K[G]$, recall that the finite set $\operatorname{supp}(z) = \{x \in G \mid a_x \neq 0\}$ is called the *support of* z.

3.2.6 Theorem: *Let G be a finite group. Assume that the following conditions are satisfied.*

H_1: *The set G is totally ordered under some relation " $\leq$ " with $1 \leq x$ for all $x \in G$.*

H_2: *There is a subset X of G such that $1 \in X$.*

H_3: *For each $y \in G - X$, $\exists\, g_y \in K[G]$ such that $1 \in (\operatorname{supp}(g_y))$ and $y\theta \leq y$ for all $\theta \in (\operatorname{supp}(g_y))$.*

Then the set $B = X \cup \{yg_y \mid y \in G - X\}$ is a K–basis of $K[G]$.

Proof: Since G is a basis of $K[G]$ and the set B does not have more than n elements, where n is the order of G, it suffices to show that B spans $K[G]$ over K. This in effect is equivalent to showing that each $y \in G - X$ is a K–linear sum of elements of B.

Let $y \in G - X$. Under the total order on G, let $G - X = \{y_1 < \cdots < y_m\}$ with $y = y_j$ for some $j \leq m$. We proceed by induction on m. Suppose $m = 1$ so that $y = y_1$ and $X = G - \{y\}$.

By H_3, we have $g_y = \sum_{\theta \in G} a_\theta \theta$ with $a_1 \neq 0$ and $a_\theta \neq 0$ implies that $y\theta \leq y$. Now we have

$$yg_y = a_1 y + \sum_{\theta \neq 1} a_\theta y\theta = a_1 y + \sum_{x \in X} a_{y^{-1}x} x \ (\text{ since } \theta \neq 1 \Rightarrow x = y\theta \neq y).$$

Thus $y = a_1^{-1} y g_y - \sum_{x \in X} b_x x$, as required. ·

Assume that $m \geq 2$ and proceed by a second induction on j. The above argument shows that if $j = 1$, the result follows since

$$1 \neq \theta \in (\text{supp}(g_{y_1})) \Rightarrow y_1 \theta < y_1 \Rightarrow y_1 \theta \in X.$$

We may assume that $j \geq 2$ and the induction hypothesis for all y_i, $i \leq j - 1$, i.e., $Y = X \cup \{y_1, \cdots, y_{j-1}\}$ is in the span of B. By H_3 again, we get that $1 \neq \theta \in (\text{supp}(g_{y_j})) \Rightarrow y_j \theta < y_j \Rightarrow y_j \theta \in Y$ and hence we find that $y_j = a_1^{-1} y_j g_{y_j} + \sum_{\theta \in Y} b_\theta \theta$ is in the span of B, as required. $\Diamond$

As an immediate consequence, we have the following.

3.2.7 Corollary: *Let M be a cyclic $K[G]$-module, in particular a simple module, generated by an element v. Assume that the following conditions are satisfied.*

H$_1$: *The set G is totally ordered under some relation " $\leq$ " with $1 \leq x$ for all $x \in G$.*

H$_2$: *There is a subset X of G such that $1 \in X$ and $Xv = \{xv \mid x \in X\}$ is linearly independent in M.*

H$_3$: *For each $y \in G - X$, $\exists \ g_y \in K[G]$ such that $g_y v = 0$, $1 \in (\text{supp}(g_y))$ and $y\theta \leq y$ for all $\theta \in (\text{supp}(g_y))$.*

Then the set $B = Xv$ is a K-basis of M and the set $B^- = \{g_y \mid y \in G - X\}$ is a K-basis of the annihilator ideal of v. ∎

3.3   Representations

Given a vector space V over a field K, we denote by $\mathrm{GL}_K(V)$ or $\mathrm{Aut}_K(V)$, the group of all K–linear automorphisms of V. We note that this is the group of all units in the ring $\mathrm{End}_K(V)$ of all K–linear endomorphisms of V. In case V is of finite dimension d over K, we identify $\mathrm{End}_K(V)$ with the ring $M_d(K)$ of all $d \times d$ matrices over K by means of a chosen K–basis of V and $\mathrm{Aut}_K(V)$ is the group $\mathrm{GL}_d(K)$ of all invertible matrices in $M_d(K)$.

3.3.1 Representations: Given a group G (finite or not) and a field K, by a *representation* of G over K, we mean a pair (V, ϱ), where V is a vector space over K and $\varrho : G \to \mathrm{Aut}_K(V)$ is a homomorphism of groups.

3.3.2 Direct sum of representations: Given two representations (V, ϱ) and (W, ϑ) of G over the same field K, by their *direct sum*, we mean the representation (L, ϖ) of G where $L = V \oplus W$ and $\varpi(x) = \varrho(x) + \vartheta(x)$, $\forall\, x \in G$. We write $\varpi = \varrho \oplus \vartheta$.

Note: Direct sum of a family $\{(V_i, \varrho_i) \mid i \in I\}$ of representations of G is defined in an obvious way and is denoted by $(\oplus_{i \in I} V_i, \oplus_{i \in I} \varrho_i)$.

3.3.3 Proposition: *A representation of G over K gives rise to a (unitary left) $K[G]$–module in a natural way and conversely. (In other words, the family of all representations of G over K is the same as the family of all (unitary left) $K[G]$–modules).*

Proof: Let (V, ϱ) be representation of G over K. Define a left scalar multiplication of $K[G]$ on V by $av = \sum_{x \in G} a_x\left(\varrho(x)\right)(v)$ for $a = \sum_{x \in G} a_x x \in K[G]$ and $v \in V$. This makes sense because by definition of $a = \sum_{x \in G} a_x x \in K[G]$, the coefficients $a_x \in K$ are all zero except for finitely many $x \in G$ and hence $\overline{\varrho}(a) \overset{\text{def}}{=} \sum_{x \in G} a_x \varrho(x)$ is a K–linear endomorphism of V (being a finite K–linear combination of the automorphisms $\varrho(x)$ of V). It is trivial to check that the map $\overline{\varrho}\colon K[G] \to \mathrm{End}_K(V)$, $a \mapsto \overline{\varrho}(a)$, is a homomorphism of K–algebras making V into a left $K[G]$–module.

Conversely, let V be a left $K[G]$–module. In particular, V is a

K–module as well, i.e., a vector space over K. But recall that giving a left $K[G]$–module structure on V is the same as giving a homomorphism of rings $\varphi : K[G] \to \text{End}_{\mathbb{Z}}(V)$ with $\varphi(1) = \text{id}_V$. Since K is a central subring of $K[G]$, it follows that $\text{Image}(\varphi) \subseteq \text{End}_K(V)$ and so φ is a homomorphism of the K–algebra $K[G]$ into $\text{End}_K(V)$. Since $G \subseteq \mathcal{U}(K[G])$ and $\varphi(\mathcal{U}(K[G])) \subseteq \mathcal{U}(\text{End}_K(V)) = \text{Aut}_K(V)$, the restriction of φ to G, say ϱ does give a homomorphism of the group G into $\text{Aut}_K(V)$ and so (V, ϱ) is a representation of G, as required. Furthermore, the left $K[G]$–module structure given by this ϱ on V coincides with the one we started with. i.e., $\varphi = \overline{\varrho}$, as required. $\Diamond$

3.3.4 Terminology: All the basic concepts and terminology in the context of $K[G]$–modules have their paralells with the same or slightly different nomenclature for the representations of G which we enlist here for ready reference. We shall use either terms without further comment.

1. Given a representation (V, ϱ) of a group G over a field K, we say that V is a *representation space* or a *G–space* or a *G–module*. We also say that G acts on V as a group of K–linear automorphisms by the action $x \cdot v = \varrho(x)(v)$, $\forall x \in G$ and $v \in V$. When there is no scope for confusion, we simply write xv for $\varrho(x)(v)$ and also just V for (V, ϱ)

The dimension of V over K is called the *dimension* or *rank* of the representation. In case V is finite dimensional over K, we say that V is a *finite dimensional representation* of G.

2. By a *G–subspace* or *G–stable subspace* or *G–submodule* W of V, we mean a $K[G]$–submodule of V, i.e., W is a vector subspace of V which is *stable* for G in the sense that $\varrho(x)(w) \in W$, $\forall x \in G$ and $w \in W$, or equivalently, the restrictions of $\varrho(x)$ to W give automorphisms of W for all $x \in G$ or what is the same, (W, ϱ) is itself a representation of G, called a *sub-representation* of V.

The notion of *quotient representation* by a sub-representation is obviously the corresponding quotient module by the submodule as $K[G]$–modules.

3. Given two representations (V, ϱ) and (W, ϑ) of G over the same field K, by a *homomorphism* of representations, or a G–map, we mean a $K[G]$–linear homomorphism of the $K[G]$–modules V and W, or equivalently, a K–linear transformation $f: V \to W$ which commutes with the actions of G on V and W, i.e., $f \circ \varrho(x) = \vartheta(x) \circ f$ for all $x \in G$, or equivalently, the following diagram commutes for all $x \in G$.

$$
\begin{array}{ccc}
V & \xrightarrow{f} & W \\
\downarrow{\varrho(x)} & & \downarrow{\vartheta(x)} \\
V & \xrightarrow{f} & W
\end{array}
$$

4. The set of all homomorphisms of representations V and W is denoted by $\mathrm{Hom}_{K[G]}(V, W)$ or simply $\mathrm{Hom}_G(V, W)$. The homomorphism theorems obviously hold for G–maps. The concepts of kernels, images, etc., for G–maps is the same as for the corresponding $K[G]$–linear maps.

5. If G is a finite group of order n such that $K[G]$ is semi–simple, i.e., "$n \neq 0$ in K", then the *averaging* process over the group G gives a K–linear projection of the vector space $\mathrm{Hom}_K(V, W)$ into the vector space $\mathrm{Hom}_G(V, W)$ for any two G–modules V and W. The averaging process, already used in the proof of (3.1.7) above, is the following.

Let $f : V \to W$ be a K–linear transformation of two G–modules (V, ϱ) and (W, ϑ). Define $\widetilde{f} : V \to W$, by

$$
\widetilde{f} = \frac{1}{n} \sum_{x \in G} (\vartheta(x))^{-1} \circ f \circ \varrho(x) = \frac{1}{n} \sum_{x \in G} \vartheta(x^{-1}) \circ f \circ \varrho(x).
$$

We note that $(\widetilde{\widetilde{f}}) = \widetilde{f}$ and in particular, $\widetilde{f} = f$, $\forall f \in \mathrm{Hom}_G(V, W)$.

3.3.5 Examples: 1. Trivial representation: Any vector space V over K can be treated as a G–module under the *trivial action* of G, namely, $xv = v$, $\forall x \in G$ and $v \in V$, i.e., under the trivial homomorphism of groups $x \mapsto 1$ of G into $\mathrm{Aut}_K(V)$. Explicitly, the scalar multiplication of $K[G]$ on V is given by $(\sum_{x \in G} a_x x)v = (\sum_{x \in G} a_x)v$. In particular, K itself can be considered as a trivial G–module, called the *trivial one dimensional representation*.

2. Regular representation: The group algebra $K[G]$ is naturally a G–module, considered as a left module over itself, called the *left* *regular representation* of G. Similarly, the *right regular representation* is simply considering $K[G]$ as a right module over itself.

When G is finite with $z = \sum_{x \in G} x \in K[G]$, the one–dimensional subspace Kz of $K[G]$ is a trivial G–submodule of $K[G]$ since $xz = z$, $\forall \, x \in G$.

3. Permutation representation: Let X be a G–set, i.e., a set on which a group G acts as a *group of bijections*. (The action of G on X permutes the elements of X). Let V be a vector space over K with X as a basis. Then V is a left $K[G]$–module in a natural way, called the *permutation representation* of G over K afforded by X.

Special cases of permutation representations are the ones given by the set $G/H = \{xH \mid x \in G\}$ of cosets of a subgroup H of G on which G acts on the left by permuting the cosets. In case $H = \{1\}$, the corresponding permutation representation is simply the left regular representation.

3.3.6 Equivalence of representations: Two representations (V, ϱ) and (W, ϑ) of G are said to be *equivalent* if they are isomorphic as $K[G]$–modules. Explicitly, this means that there is an isomorphism $f \colon V \to W$, such that $f \circ \varrho(x) = \vartheta(x) \circ f$, $\forall \, x \in G$.

3.3.7 Irreducible representations: A representation V of G is called *irreducible* or *simple* if $V \neq (0)$ and the only G–submodules of V are (0) and V, i.e., V is simple as a $K[G]$–module.

Note: Any 1–dimensional representation is obviously irreducible but not conversely. However, a trivial representation is irreducible $\iff$ it is 1–dimensional (since any vector subspace is then a G–submodule).

3.3.8 Complete reducibility: A representation V of G is said to be *completely reducible* if V is semi–simple as a $K[G]$–module, i.e., every G–submodule of V is a direct summand or equivalently, V is a direct sum of a family of irreducible representations of G.

3.3.9 Remark: A representation of a group *need not* be completely reducible. For example, the left regular representation $K[G]$ of G over a field of characteristic p with p dividing the order of G is *not* semi-simple (3.1.7). The facts we summarised in (3.2.2) and (3.2.3) can be reformulated in the language of representations as follows.

3.3.10 Theorem: *Let G be a finite group of order n. Let K be an algebraically closed field such that $K[G]$ is semi–simple, (for example, $K = \mathbb{C}$). Write $K[G] = \prod_{i=1}^{h} M_{m_i}(K)$ where h is the number of conjugacy classes of G. Then we have the following.*

1. *There are exactly h mutually inequivalent irreducible representations V_i of G such that any irreducible representation of G is equivalent to one of them, i.e., $\{V_i \mid 1 \leq i \leq h\}$ is a complete set of mutually inequivalent irreducible representations of G. In particular, the one-dimensional trivial representation K, being irreducible, is equivalent to one of them say V_1.*

2. *The representation V_i can be taken to be the space of all matrices in $M_{m_i}(K)$ all of whose columns but the first are zero. In particular, $\dim_K(V_i) = m_i$, $1 \leq i \leq h$, with $m_1 = 1$.*

3. *We have $n = \sum_{i=1}^{h} m_i^2 = 1 + \sum_{i=2}^{h} m_i^2$. Furthermore, $n = h \Leftrightarrow G$ is abelian $\Leftrightarrow m_i = 1$, $\forall\, i$, $1 \leq i \leq h$. In particular, all irreducible representations of an abelian group are one-dimensional.*

4. *Every representation of G is completely reducible and decomposes into a direct sum of isotypical components of type V_i, $1 \leq i \leq h$.*

5. *Given a finite dimensional representation V of G, there exist unique non–negative integers d_i, called the decomposition numbers of V, such that $V \approx V_1^{\oplus d_1} \oplus \cdots \oplus V_h^{\oplus d_h}$ and hence $d = \dim_K(V) = \sum_{i=1}^{h} d_i m_i$. (Uniqueness is that of the isotypical components of V).*

6. *The integer d_i is called the* multiplicity *with which the irreducible component V_i occurs in V.*

7. *The decomposition numbers of the regular representation $K[G]$ are $d_i = m_i$, $\forall\, i$, $1 \leq i \leq h$.*

8. *A representation each of whose decomposition numbers is 0 or 1 is said to be a* multiplicity free *representation.*

3.3.11 Remarks: For any irreducible representation V of G over K, we have the following.

1. $\mathrm{End}_G(V) = K$ (since by Schur's lemma, $\mathrm{End}_G(V)$ is a division subring of $\mathrm{End}_K(V)$) and hence finite dimensional over K but K is algebraically closed, implying the result by (Ex.(2.9.15) above).

2. For each element $x \in Z(G)$ (the centre of G), the homothecy defined by x on V is a scalar multiplication by some non–zero element of K (because the K–linear automorphism ℓ_x of V is G–linear too and so ℓ_x is in $\mathrm{End}_G(V) = K$).

3. In fact, for each conjugacy class C of G, if $z = \sum_{x \in C} x$, then the homothecy ℓ_z on V is a scalar multiplication by some element of K.

3.3.12 Invariants: Given a representation V of G, a vector $v \in V$ is called an *invariant* or a *fixed point* for G if $gv = v$ for all $g \in G$.

1. The set of all G–invariants in V is denoted by V^G which is a subspace of V, called the *space of invariants*.

2. In case $V^G = (0)$, we say that V is a *fixed point free*. (A non–trivial irreducible representation is fixed point free but not vice–versa).

3. It is obvious that V^G is the *largest trivial G–submodule* of V and its dimension is simply the multiplicity with which the trivial irreducible representation V_1 of G occurs in V. (In particular, $(K[G])^G = V_1 = Kz$ where $z = \sum_{x \in G} x$). ■

3.4 Characters of Representations

As usual, we assume that G is a finite group of order n and K is an algebraically closed field such that $K[G]$ is semi–simple, for instance $K = \mathbb{C}$. Unless specified otherwise, all representations of G considered hereafter are assumed to be finite dimensional over K.

3.4.1 Character: Given a representation (V, ϱ) of G over K, the K–valued set map on G, $x \mapsto \mathrm{trace}(\varrho(x))$, is called the *trace function* or the *character* of (V, ϱ) and is denoted by χ_ϱ or χ_V or simply χ when there is no scope for confusion.

3.4.2 Remarks: 1. Choosing a K-basis of V, we can represent each $\varrho(x)$, $x \in G$, by a $d \times d$ matrix $\varrho(x) = (\varrho(x)_{ij})$ over K where $d = \dim_K(V)$ and find that $\chi(x) = \operatorname{trace}(\varrho(x)) = \operatorname{trace}(\varrho(x)_{ij}) = \sum_{i=1}^{d} \varrho(x)_{ii}$ is independent of the basis chosen for V.
2. For each $x \in G$, we can find a K-basis of V, depending on x, with respect to which the matrix $(\varrho(x)_{ij})$ of $\varrho(x)$ is diagonal, i.e., each of the automorphisms $\varrho(x)$ of V can be *diagonalised* (though not simultaneously).

To see this, fix an $x \in G$ and consider the cyclic subgroup $< x >$ generated by x. Since V is a G-module, by restriction, V is also a $< x >$-module. On the other hand, since $K[G]$ is semi-simple, so is $K[< x >]$ and hence V is a semi-simple $K[< x >]$-module which implies that V is a direct sum of simple $K[< x >]$-modules. But then by (3.3.10)(3) above, each simple $K[< x >]$-module is one-dimensional since $< x >$ is abelian. Thus $\exists\, v_i \in V$ such that $V = \oplus_{i=1}^{d} K v_i$ with $\varrho(x)(v_i) = a_i v_i$ for some $a_i \in K$, $1 \le i \le d$. Hence $\varrho(x) = \operatorname{diag}(a_1, \cdots, a_d)$ with respect to the basis $\{v_1, \cdots, v_d\}$ of V.

Note: The diagonalisation of $\varrho(x)$ is *not* true if K is not algebraically closed even if $K[G]$ is semi-simple. For instance, fix an *odd* prime number p and take $K = \mathbb{Q}$ and $G = \mathbb{Z}/p\mathbb{Z}$, the cyclic group of order p generated, say by x and $V = \mathbb{Q}[\mathbb{Z}/p\mathbb{Z}]$, the regular representation which is semi-simple. Using the facts about $\mathbb{Q}[\mathbb{Z}/p\mathbb{Z}]$ from (3.2.5), we find that the left multiplication by x cannot be diagonalised.

3.4.3 Some Examples: 1. Let (V, ρ) be the trivial representation of dimension d. Then $\rho(x) = \operatorname{id}_V$, $\forall\, x \in G$ and hence $\chi_\rho(x) = d \cdot 1 \in K$, i.e., χ_ρ is constant. In particular, $\chi_K \equiv 1$ for the trivial 1-dimensional representation.

2. Let $(K[G], \operatorname{reg})$ be the regular representation. Then

$$\chi_{\mathbf{reg}}(x) = \begin{cases} n \cdot 1 & \text{if } x = 1, \\ 0 & \text{if } x \ne 1. \end{cases}$$

This follows from the fact that the matrix of left multiplication by any $x \ne 1$ with respect to the basis G for $K[G]$ has zeros on the main diagonal.

This is a particular case of the following (corresponding to the trivial subgroup of G).

3. Let H be a subgroup of G and $(K^{[G:H]}, \text{perm}_H)$ be the permutation representation of G on the vector space with G/H as a basis (3.3.5)(3). Then the character of this representation is given as follows.

3.4.4 Proposition: *For $x \in G$, let C_x be the conjugacy class through x and $C_G(x) = \{y \in G | xy = yx\}$ be the centraliser of x in G. Then we have*

$$\chi_{\text{perm}_H}(x) = [G : H]\frac{|C_x \cap H|}{|C_x|}$$

where $|X|$ denotes the cardinality of X.

Proof: Let $\{x_1, \cdots, x_m\}$ be a complete set of coset representatives for H where $m = [G : H]$. Let ℓ be the order of H. It is trivial to see that we have

$$\begin{aligned}
\chi_{\text{perm}_H}(x) &= |\{j \mid xx_jH = x_jH\}| = \left|\{\, j \mid x_j^{-1}xx_j \in H\}\right| \\
&= \frac{|\{y \in G \mid y^{-1}xy \in H\}|}{\ell} = \frac{|C_x \cap H||C_G(x)|}{\ell} \\
&= \frac{n\ |C_x \cap H|}{\ell\ |C_x|} \quad (\text{since } C_x \overset{\text{bij}}{\approx} G/C_G(x)) \\
&= [G : H]\frac{|C_x \cap H|}{|C_x|}, \quad \text{as required.} \qquad \diamond
\end{aligned}$$

3.4.5 Remark: Equivalent representations have the same character but *not* conversely (*unless* **(i)** the representations are one–dimensional ((3.5.3) below) or more generally **(ii)** irreducible ((3.6.5) below) or **(iii)** K is of characteristic 0 ((3.7.2) below).

For, let (V, ϱ) and (W, ϑ) be equivalent, say $f\colon V \overset{\sim}{\to} W$ be an isomorphism. We know that $\vartheta(x) \circ f = f \circ \varrho(x)$, $\forall\, x \in G$, i.e., $\vartheta(x) = f \circ \varrho(x) \circ f^{-1}$ and hence for all $x \in G$, we have

$$\chi_\vartheta(x) = \text{trace}(\vartheta(x)) = \text{trace}(f \circ \varrho(x) \circ f^{-1}) = \text{trace}(\varrho(x)) = \chi_\varrho(x).$$

As for the failure of the converse, suppose K is of characteristic $p \neq 0$. Take the trivial representations on $V = K^p$ and $W = K^{2p}$ which are obviously non–equivalent for dimension reasons but $\chi_V \equiv p \cdot 1 = 0 = 2p \cdot 1 \equiv \chi_W$.

3.4.6 Proposition: *Let (V, ϱ) and (W, ϑ) be two irreducible representations of G. Let $f \in \mathrm{Hom}_K(V, W)$ and $\widetilde{f} \in \mathrm{Hom}_G(V, W)$ be the average of f over G ((3.3.4)(5) above). Then we have the following.*
1. $\widetilde{f} \equiv 0$ *if V is not equivalent to W.*
2. *If $V = W$ and $\varrho = \vartheta$, then $\widetilde{f}$ is a scalar multiplication on V and* $\mathrm{trace}(\widetilde{f}) = \mathrm{trace}(f)$.

Proof: Since V and W are irreducible and $\widetilde{f} \in \mathrm{Hom}_G(V, W)$, $\widetilde{f} = 0$ if V is not G–isomorphic to W (by Schur's lemma). On the other hand, if $V = W$ and $\varrho = \vartheta$, we have $\mathrm{End}_G(V) = K$ (by (3.3.11) above) since K is algebraically closed. Hence $\widetilde{f}$ is a *homothecy* (i.e., multiplication) by a scalar in K. In other words, we have $\widetilde{f} = a(\mathrm{id}_V)$ for some $a \in K$. Finally, we have

$$\mathrm{trace}(\widetilde{f}) = \frac{1}{n} \sum_{x \in G} \mathrm{trace}\left(\varrho(x^{-1}) \circ f \circ \varrho(x)\right)$$

$$= \frac{1}{n} \, n \, \mathrm{trace}(f) = \mathrm{trace}(f). \qquad \Diamond$$

3.4.7 Corollary: *If V is an irreducible representation of G, then the dimension of V (over K) is not a multiple of the characteristic of K. (This is not true if $K[G]$ is not semi–simple. Give an example.)*

For, since V is not zero, we can take an $f \in \mathrm{End}_K(V)$ of trace 1. Now $\widetilde{f} \neq 0$ and is a homothecy by a scalar $a \in K$, i.e., $\widetilde{f} = a(\mathrm{id}_V) \neq 0$ and hence $1 = \mathrm{trace}(f) = \mathrm{trace}(\widetilde{f}) = d \, a \Rightarrow d \cdot 1 \neq 0$ in K, where $d = \dim_K(V)$, as required. $\qquad \Diamond$

3.4.8 Irreducible characters: The character of an irreducible representation is called an *irreducible character* or a *simple character*.

Note: Irreducible characters are non–zero K–valued functions on G because (by (3.4.7) above) we have $\chi_V(1) = d \cdot 1 \neq 0$ in K for any irreducible representation V of G over K (where $d = \dim_K(V)$).

3.4.9 Theorem: **(i)** *The character of a finite direct sum of representations is the sum of their characters.*

(ii) *The character of any representation is a sum of irreducible characters, called its* irreducible components.

Proof: Easy verification. ◇

3.4.10 Corollary: $\chi_{\text{reg}} = \sum_{i=1}^{r} m_i \, \chi_{V_i}$, *where* $m_i = dim_K(V_i)$, $1 \leq i \leq r$, *are also the decomposition numbers of* $(K[G], \text{reg})$. *In other words, an irreducible representation occurs in the regular representation* $K[G]$ *exactly as many times as its dimension.* ∎

3.5 Group Characters

3.5.1 Group character: By a *group character* or a *character of a group* G, we mean a one–dimensional representation of G.

Note: Given two group characters (K, ρ_1) and (K, ρ_2), it is trivial to check that their product $(K, \rho_1\rho_2)$ is again a character of G where $(\rho_1\rho_2)(x) = \rho_1(x)\rho_2(x) = (\rho_2\rho_1)(x)$, $\forall \, x \in G$.

The notion of a group character (or character of a group) as against the character of a representation is justified by the following.

3.5.2 Proposition: *The character* χ_ρ *of a group character* (K, ρ) *is equal to* ρ *as maps on* G *with values in* $K^\star$.

Proof: Trivial. ◇

3.5.3 Corollary: *Two characters of a group are* equal *if and only if they are* equivalent.

3.5.4 Remark: The set $\text{Hom}_{\text{groups}}(G, K^\star)$ of characters of a group G is denoted by $\widetilde{G}$ and it is trivial to see that $\widetilde{G}$ is an *abelian group* under multiplication of characters, called the *character group* of G.

We shall now determine the size of the group $\widetilde{G}$. To do this, we need to consider *representations of quotient groups*. Let H be a *normal* subgroup of G and $\eta_H : G \to G/H$ be the natural homomorphism. Then we have the following.

3.5.5 Proposition: *Given a representation (V, ϱ) of G, ϱ gives rise (or goes down) to a representation $(V, \overline{\varrho})$ of G/H if and only if ϱ is trivial on H, i.e., $H \subseteq \mathrm{Ker}\varrho$ and in that case $\varrho = \overline{\varrho} \circ \eta_H$.*

Proof: This is an immediate consequence of the epimorphism theorem for groups.   ◇

Recall that for a group G, the quotient group $G^{\mathrm{ab}} = G/G^{(1)}$ is called the *abelianiser* of G where $G^{(1)}$ is the *commutator subgroup* of G. It is the largest abelian quotient of G since $G^{(1)}$ is the smallest normal subgroup H of G such that G/H is abelian.

3.5.6 Theorem: (i) *The character group $\widetilde{G}$ of a group G is isomorphic to the character group $\widetilde{G^{\mathrm{ab}}}$ of the abelianiser G^{ab} of G.*
(ii) *We have $G \approx \widetilde{G}$ if and only if G is abelian.*

Proof: Suppose (K, ρ) is a character of G. Since $G/\mathrm{Ker}(\rho) \approx \mathrm{Image}(\rho)$ which is a subgroup of the abelian group $K^{\star}$, we get that $G^{(1)} \subseteq \mathrm{Ker}\ \rho$ and hence ρ induces a character $\overline{\rho}$ of G^{ab} and conversely any character of G^{ab} gives rise to one of G. Thus the mapping $\rho \mapsto \overline{\rho}$ can be easily seen to be an isomorphism of the groups $\widetilde{G}$ and $\widetilde{G^{\mathrm{ab}}}$, as required.

(ii) This follows from the fact that any irreducible representation of an abelian group is one–dimensional (i.e., a character) (by (3.3.10)(3) above) and there are exactly n *distinct* characters of G. We know that the $\left|\widetilde{G}\right| \leq |G|$ with equality if and only if G is abelian. Now look at the natural evaluation map, $e\colon G \to \widetilde{\widetilde{G}}$, $a \mapsto e_a$ where $e_a(\chi) = \chi(a)$ for all $\chi \in \widetilde{G}$ and $a \in G$, which is obviously a homomorphism of groups if G is abelian. If $\mathrm{Ker}(e) = H$, then it follows that every character of G goes down to a character of G/H, and hence $\widetilde{G} = \widetilde{(G/H)}$ implying that, if $H \neq \{1\}$, $\left|\widetilde{G}\right| \leq \left|\widetilde{G/H}\right| < |G|$ which would be a contradiction. Hence e is a monomorphism between two groups of the same order and hence an isomorphism, as required.   ◇

3.5.7 Corollary: *The number of distinct characters of G is the index of the commutator subgroup $G^{(1)}$ of G.*   ■

3.6   Orthogonality Relations

Let G and K be as in § 3.4 above.

3.6.1   Class functions: A K–valued set map $f : G \to K$ on G is called a *class function* if it is constant on each conjugacy class of G.

The set of all class functions on G is a vector space over K and is denoted by $\mathcal{C}_K(G)$.

3.6.2   Examples: Constant functions and characters of representations of G are class functions on G.

3.6.3   Proposition: *The set $\mathcal{F}_K(G)$ of all K–valued set maps on G is a vector space of dimension n ($=$ order of G) over K containing $\mathcal{C}_K(G)$ as a subspace of dimension h where h is the number of conjugacy classes of G.*

Proof: The set of all characteristic functions ψ_x of elements x of G is a K–basis of $\mathcal{F}_K(G)$ and hence of dimension n, the order of G. Recall that the characteristic function ψ_x of an element $x \in G$ is the K–valued map on G given by

$$\psi_x(y) = \begin{cases} 1 & \text{if } y = x, \\ 0 & \text{if } y \neq x. \end{cases}$$

Likewise, the set of all characteristic functions ψ_C of conjugacy classes C of G is a K–basis of $\mathcal{C}_K(G)$ where

$$\psi_C(x) = \begin{cases} 1 & \text{if } x \in C, \\ 0 & \text{if } x \notin C. \end{cases} \qquad \Diamond$$

For simplicity of notation, in the rest of this section, we write $\mathcal{F}$ $= \mathcal{F}_K(G)$ and $\mathcal{C} = \mathcal{C}_K(G)$.

3.6.4   The inner poduct $\langle \, , \, \rangle_G$: We introduce an *inner product* $\langle \, , \, \rangle_G = \langle \, , \, \rangle$ on $\mathcal{F}$ as follows. For $f, g \in \mathcal{F}$, let

$$\langle f \, , \, g \rangle = \frac{1}{n} \sum_{x \in G} f(x^{-1}) g(x) = \frac{1}{n} \sum_{x \in G} g((x^{-1})^{-1}) f(x^{-1}) = \langle g \, , \, f \rangle .$$

Thus $\langle \, , \, \rangle$ is a symmetric K–bilinear form on $\mathcal{F} \times \mathcal{F}$.

3.6.5 Theorem (Orthogonality relations–First kind): *Let* (V, ϱ)
and (V', ϱ') *be two irreducible representations of* G *with their charac-*
ters $\chi = \chi_\varrho$ *and* $\chi' = \chi_{\varrho'}$ *respectively. Then we have the following.*

$$\langle \chi \,,\, \chi' \rangle = \begin{cases} 0 & \text{if } V \text{ is not equivalent to } V', \\ 1 & \text{if } V \text{ is equivalent to } V'. \end{cases}$$

Hence, the set $\{\chi_{V_i} \mid 1 \leq i \leq h\}$ *of characters of a complete set of*
mutually inequivalent irreducible representations $\{(V_i, \varrho_i) \mid 1 \leq i \leq h\}$
of G *is an orthonormal basis of the space* $\mathcal{C}$ *of class functions on* G
where h *is the number of conjugacy classes of* G.

Proof: Let $f \in \text{Hom}_K(V, V')$. With respect to some chosen
K–bases for V and V', write the matrices

$$\varrho(x) = \left(\varrho_{ij}(x) \right), \quad \varrho'(x) = \left(\varrho'_{kl}(x) \right) \text{ and } f(x) = \left(f_{qt}(x) \right)$$

for $1 \leq i, j, t \leq d$ and $1 \leq k, l, q \leq d'$, where $d = \dim_K(V)$ and $d' = \dim_K(V')$. If $\widetilde{f}$ is the average of f over G ((3.3.4)(5) above), then it
is easy to see that the matrix of $\widetilde{f}$ is given by

$$\widetilde{f} = \left((\widetilde{f})_{qt} \right) \quad \text{where} \quad (\widetilde{f})_{qt} = \frac{1}{n} \sum_{x,l,i} \varrho'_{ql}(x^{-1}) f_{li} \varrho_{it}(x).$$

If V is not equivalent to V', we get that $\widetilde{f} \equiv 0$, by (3.4.6) above.
Thus $(\widetilde{f})_{qt} = 0$, $\forall\, q, t$ and $\forall\, f \in \text{Hom}_K(V, V')$ and hence making a
suitable choice of f, viz., $f_{qt} = 1$, $\forall\, q, t$, we conclude that

$(\star) \qquad \left\langle \varrho'_{ql}, \varrho_{it} \right\rangle = \frac{1}{n} \sum_{x \in G} \varrho'_{ql}(x^{-1}) \varrho_{it}(x) = 0, \; \forall\, q, l, i \text{ and } t.$

Now we have

$$\begin{aligned} \langle \chi' \,,\, \chi \rangle &= \frac{1}{n} \sum_{x \in G} \chi'(x^{-1}) \chi(x) = \frac{1}{n} \sum_{x \in G} \left[\sum_{q,t} \varrho'_{qq}(x^{-1}) \varrho_{tt}(x) \right] \\ &= \sum_{q,t} \left[\frac{1}{n} \sum_{x \in G} \varrho'_{qq}(x^{-1}) \varrho_{tt}(x) \right] = \sum_{q,t} \left\langle \varrho'_{qq} \,,\, \varrho_{tt} \right\rangle = 0. \end{aligned}$$

To see the second part, since equivalent representations have the same
character, we may assume that $V = V'$ and $\varrho = \varrho'$. Now, again

by (3.4.6) above, we know that for all $f \in \text{End}_K(V)$, $\widetilde{f}$ is a scalar multiplication on V by some $a = a_f \in K$ and $\text{trace}(f) = \text{trace}(\widetilde{f}) = a/d$, i.e.,

$$(\widetilde{f})_{qt} = \begin{cases} 0 & \text{if } q \neq t, \\ a/d & \text{if } q = t. \end{cases}$$

where $a = \sum_i f_{ii} = \text{trace}(f)$. Since V is irreducible, we get that $d \neq 0$ in K (by (3.4.6) above).

As before, by making a suitable choice of the f_{li}'s, it can be seen easily that

$$(\star\star) \quad \langle \varrho_{ql}, \varrho_{it} \rangle = \frac{1}{n} \sum_{x \in G} \varrho_{ql}(x^{-1}) \varrho_{it}(x) = \begin{cases} 0 & \text{if } q \neq t \text{ or } l \neq i, \\ 1/d & \text{if } q = t \text{ and } l = i. \end{cases}$$

In particular, we have

$$(\star\star\star) \qquad \langle \varrho_{qq}, \varrho_{tt} \rangle = \begin{cases} 0 & \text{if } q \neq t, \\ 1/d & \text{if } q = t. \end{cases}$$

Therefore this gives that

$$\begin{aligned} \langle \chi, \chi \rangle &= \frac{1}{n} \sum_{x \in G} \chi(x^{-1}) \chi(x) = \frac{1}{n} \sum_{x \in G} \left[\sum_{q,t=1}^{d} \varrho_{qq}(x^{-1}) \varrho_{tt}(x) \right] \\ &= \sum_{q=1}^{d} \left[\frac{1}{n} \sum_{x \in G} \varrho_{qq}(x^{-1}) \varrho_{qq}(x) \right] = \sum_{q=1}^{d} \langle \varrho_{qq}, \varrho_{qq} \rangle \\ &= d\frac{1}{d} = 1, \quad \text{as required.} \qquad \diamond \end{aligned}$$

3.6.6 Corollary: *A class function* $f \in C$ *can be uniquely expressed as* $f = \sum_{i=1}^{h} \langle f, \chi_i \rangle \chi_i$. *In particular, if* χ *is the character of a representation whose decomposition numbers are* $\{d_i\}_{i=1}^{h}$, *then we have*

$$\chi(x) = \sum_{i=1}^{h} d_i \chi_i(x) = \sum_{i=1}^{h} \langle \chi, \chi_i \rangle \chi_i(x), \ \forall \ x \in G.$$

Remark: The relations $(\star)$ and their special cases $(\star\star)$ and $(\star\star\star)$ above, are *also* called the *orthogonality relations of the first kind* satisfied by the *matrix coefficient functions* of an irreducible representations of G.

Exercises: (1) For any irreducible representation (V, ϱ) of G, show that the set $\varrho(G) = \{\varrho(x) \mid x \in G\}$ of automorphisms of V spans the vector space $\text{End}_K(V)$. (Use the orthogonality relations $(\star\star)$ above, if the dimension of the span of $\varrho(G)$ is short of d^2 where $d = \dim_K(V)$).

(2)   Theorem(Frobenius–Schur): Let $\{(V_\ell, \varrho_\ell) \mid 1 \leq \ell \leq h\}$ be a set of mutually inequivalent irreducible representations of G. Then the set of matrix coefficient functions on G, namely,

$$\{\varrho_{\ell jk} \mid 1 \leq j, k \leq \dim_K(V_\ell), \ 1 \leq \ell \leq h\}$$

is linearly independent. (Cf. [10], Chapter IV.)   ◇

3.6.7 Character table: The character values of *all representations* of G are known once we know the table below, namely, the character values of *all* the distinct *irreducible characters*, called the *character table* of the group G.

Let $\{x_i\}_{i=1}^h$ be a *complete set of representatives* of the conjugacy classes of G.

Character Table of G

	x_1	$\cdots$	x_j	$\cdots$	x_h
χ_1	$\chi_1(x_1)$	$\cdots$	$\chi_1(x_j)$	$\cdots$	$\chi_1(x_h)$
χ_2	$\chi_2(x_1)$	$\cdots$	$\chi_2(x_j)$	$\cdots$	$\chi_2(x_h)$
$\vdots$			$\ddots$		
χ_i	$\chi_i(x_1)$	$\cdots$	$\chi_i(x_j)$	$\cdots$	$\chi_i(x_h)$
$\vdots$			$\ddots$		
χ_h	$\chi_h(x_1)$	$\cdots$	$\chi_h(x_j)$	$\cdots$	$\chi_h(x_h)$

The following gives another kind of orthogonality relations between the irreducible characters with respect to their values on a pair of conjugacy classes.

3.6.8 Theorem (Orthogonality relations–Second kind): *Let* $\{C_i \mid 1 \leq i \leq h\}$ *be the set of all conjugacy classes of G with a* *set of representatives* $\{x_i \in C_i \mid 1 \le i \le h\}$. *Let* $h_i = |C_i|$. *Then we have* $\sum_{i=1}^{h} \chi_i(x_j)\chi_i(x_k^{-1}) = (n/h_j)\delta_{j,k}$, *where* δ *is the Kronecker's delta function. In particular, we have*

$$\sum_{i=1}^{h} \chi_i(x)\chi_i(x^{-1}) = \frac{n}{h_x} = |C_G(x)|, \ \forall \ x \in G.$$

Proof: By (3.6.5) above, we have

$$\delta_{j,k} = \langle \chi_j, \chi_k \rangle = \frac{1}{n}\sum_{x \in G} \chi_j(x^{-1})\chi_k(x) = \frac{1}{n}\sum_{i=1}^{h} h_k \chi_j(x_i^{-1})\chi_k(x_i).$$

This means that $AB = I_h$ (the identity matrix of order h) where

$$A = \left(\chi_j(x_i^{-1})\right) \quad \text{and} \quad B = \frac{1}{n}\left(h_p \chi_p(x_q)\right).$$

But then we must have $BA = I_h$ implying the stated identities.   ■

3.7   Ordinary and Modular Representations

3.7.1 Ordinary and modular representations: Representations of a group G over a field K are called *ordinary* representations if the characteristic of K is zero, otherwise, they are called *modular* or *p–modular* representations where $p =$ Char K. Similar terminology is used for the characters of representations.

Ordinary representations have the following special properties.

3.7.2 Theorem: *Let K be algebraically closed and of characteristic 0. Then we have the following.*
1. *Two representations of G over K are equivalent if and only if their characters are equal, i.e., an ordinary representation is determined by its character.*
2. *A representation is irreducible* $\iff$ *its character is of norm 1.*
3. *The additive subgroup $\chi_K(G)$ of $C_K(G)$ generated by the characters of representations of G is a free abelian group of rank h with the set of irreducible characters as a basis, where h is the number of conjugacy classes of G.*

Proof: (1) Let $V = V_1^{\oplus d_1} \oplus \cdots \oplus V_h^{\oplus d_h}$ be a $K[G]$–module with its decomposition numbers $d_i \in \mathbb{Z}^+$, $1 \le i \le h$. We have $\chi_V = \sum_{i=1}^h d_i \chi_i$. Similarly, if $W = V_1^{\oplus e_1} \oplus \cdots \oplus V_h^{\oplus e_h}$ is another $K[G]$–module with its decomposition numbers $e_i \in \mathbb{Z}^+$, $1 \le i \le h$, then $\chi_W = \sum_{i=1}^h e_i \chi_i$. Now $\chi_V = \chi_W \Rightarrow d_i = e_i$, $\forall\, i$ in K by the linear independence of of the χ_i's. Hence $d_i = e_i$ in $\mathbb{Z}$ since K is of characteristic 0. Hence $V \approx W$, as required.

If V is irreducible, we have already seen that $\langle \chi_V, \chi_V \rangle = 1$ whatever be the characteristic of K. On the other hand, if $\langle \chi_V, \chi_V \rangle = 1 = \sum_{i=1}^h d_i^2$, then $d_i = 1$ for some i and $d_j = 0$, $\forall\, j \neq i$ since characteristic of K is 0. Hence $V \approx V_i$ is irreducible, as required.

The last statement is obvious.   ◇

3.7.3   Virtual characters: Any element $\chi \in \chi(G) = \chi_K(G)$ is called a *virtual character* of G.

3.7.4   Remark: If $\mathrm{Char} K = 0$, a virtual character $\chi \in \chi(G)$ is in fact a *genuine character*, i.e., the character of a representation if and only if χ is a non–negative integral combination of the irreducible characters. Secondly, a virtual character of norm 1 is an irreducible character or the negative of one such.

3.7.5   Remark: If the characteristic of K is *non–zero*, then the first two statements of (3.7.2) above, are false. For example, take $W = V^{\oplus(p+1)}$ where V is an irreducible representation and p is the characteristic of K. Then we have $\chi_W = (p+1)\chi_V = \chi_V$ and hence χ_W is of norm 1 as well but W is *neither* irreducible *nor* determined by its character.

3.7.6 Remark: Given a finite group G, in order to describe the representations of G over an algebraically closed field K such that $K[G]$ is semi–simple, (in particular, to describe the ordinary representations of G), one needs to do four things, namely–
1. *Determine the set of conjugacy classes of G.*
2. *Construct irreducible representations V_C of G, parametrised by the set of conjugacy class C of G, in such a way that the following holds.*

3. *For different conjugacy classes* C *and* C'*, the representations* V_C *and* $V_{C'}$ *are not equivalent.*
4. *Determine the dimension of* V_C *over* K.

Note: We shall construct the ordinary irreducible representations of the *Symmetric group* S_n (of permutations on n symbols) in Chapter 5 below. The case of some important subgroups of S_{2n}, the so called *Hyperoctahedral groups* B_n and D_n, will be considered in Chapters 7 and 8 respectively. The *alternating subgroup* A_n of S_n will be considered in Chapter 6. The techniques of *restriction* and *induction*, to be developed in Chapter 4, will be required for A_n and D_n. The group D_n is a subgroup of B_n of index 2 and the considerations for the case (B_n, D_n) are almost similar to that of (S_n, A_n).

While the existence of a complete set $\mathrm{Irr}_K(G)$ of mutually inequivalent ordinary irreducible representations of G, parametrised by the set of conjugacy classes of G is clear, an explicit construction or realisation of the same is not easy. However, even when a realisation is possible, there is no natural way of constructing an irreducible representation associated to a given conjugacy class of G, as we shall see in the cases of A_n and D_n. Nevertheless, in the cases of S_n and B_n, there are several explicit methods of constructing them, each one being described depending on a conjugacy class. ∎

3.8 Integrality of Complex Characters

In this section, we assume that the base field is the field of Complex numbers $\mathbb{C}$. We shall study some special properties enjoyed by the characters of representations over the field $\mathbb{C}$. We call these representations the *complex representations* and their characters the *complex characters*.

3.8.1 Proposition: *Let* (V, ϱ) *be a complex representation of* G *with its character* χ*. Then we have the following.*
1. *For all* $x \in G, \chi(x)$ *is a sum of* n^{th}*-roots of unity where* n *is the order of* G *and*
2. $\chi(x^{-1}) = \overline{\chi(x)}$*, the complex conjugate of* $\chi(x)$*.*

Proof: For $x \in G$, by (3.4.2) above, diagonalise $\varrho(x)$ with respect to a suitable basis of V, say $\varrho(x) = \text{diag}(a_1, \cdots, a_d)$ where $d = \dim_{\mathbb{C}}(V)$. Since ϱ is a homomorphism of groups, we have $\varrho\,(1) = \varrho\,(x^m) = (\varrho\,(x))^m = \text{diag}(a_1{}^m, \cdots, a_d{}^m) = \text{id}_V$, where $m = \text{ord}(x)$ and hence $a_i{}^m = 1$ for all i and hence $\chi(x) = \sum_{i=1}^{d} a_i$ is a sum of mth-roots (and hence n^{th}-roots) of unity (since $m \mid n$), as required.

The last is immediate since for a complex root a of 1, we have $a\bar{a} = |\,a\,|^2 = |\,a\,| = 1$ and so $\varrho(x)\overline{\varrho(x)} = \text{diag}(a_1\overline{a_1}, \cdots, a_d\overline{a_d}) = \text{id}_V$. $\lozenge$

3.8.2 Remark: The inner product (3.6.4) $\langle\ ,\ \rangle$ defined on $\mathcal{F}_{\mathbb{C}}(G)$ can also be taken as

$$\langle f\ ,\ g \rangle = \frac{1}{n} \sum_{x \in G} f(x^{-1})g(x) = \frac{1}{n} \sum_{x \in G} \overline{f(x)}g(x) = \overline{\langle g\ ,\ f \rangle}$$

which is *not* symmetric but *conjugate symmetric* and the set of irreducible characters of G is still an orthonormal basis for $\mathcal{C}_{\mathbb{C}}(G)$.

3.8.3 Algebraic integer: A complex number λ is called an *algebraic integer*, (or said to be *integral* over $\mathbb{Z}$), if it is a root of a *monic polynomial* with integer coefficients, i.e., $\lambda^r + n_1\lambda^{r-1} + \cdots + n_{r-1}\lambda + n_r = 0$, for some integers $n_i \in \mathbb{Z}$ and $r \in \mathbb{N}$.

3.8.4 Examples: Usual integers and their n^{th} roots, in particular, roots of unity are all algebraic integers.

3.8.5 Proposition: *A rational number is an algebraic integer if and only if it is an integer.*

Proof: Let $\lambda = a/b \in \mathbb{Q}$ with $\gcd(a, b) = 1$ be an algebraic integer, say $\lambda^r + n_1\lambda^{r-1} + \cdots + n_{r-1}\lambda + n_r = 0$, for some integers $n_i \in \mathbb{Z}$ and $r \in \mathbb{N}$. On substitution, we get

$$a^r + n_1 a^{r-1}b + \cdots + n_{r-1}ab^{r-1} + n_r b^r = 0 \Rightarrow a^r = bb' \text{ for some } b' \in \mathbb{Z}.$$

If b is not a unit in $\mathbb{Z}$, take a prime divisor p of b. Now $p \mid bb'$, i.e., $p \mid a^r \Rightarrow p \mid a$ which is a contradiction and hence $b = \pm 1$. $\lozenge$

3.8.6   Theorem: *For a complex number λ, the following are equivalent.*

1. λ *is an algebraic integer.*

2. *The subring* $\mathbb{Z}[\lambda]$ *of* $\mathbb{C}$ *generated by* $\mathbb{Z}$ *and* λ *is finitely generated as an abelian group, i.e., Noetherian as a* $\mathbb{Z}$*-module.*

Proof: $(1) \Rightarrow (2)$: Let λ satisfy a monic polynomial $\lambda^r + n_1\lambda^{r-1} + \cdots + n_{r-1}\lambda + n_r = 0$, for some integers $n_i \in \mathbb{Z}$ and $r \in \mathbb{N}$. Note then that $\lambda^r = -(\sum_{i=1}^{r} n_i\lambda^{r-i}) \in \sum_{i=0}^{r-1} \mathbb{Z}\lambda^i (\subseteq \mathbb{Z}[\lambda])$. Now on repeatedly substituting for λ^r, we have

$$\mathbb{Z}[\lambda] = \{\sum_{i=0}^{s} m_i\lambda^i, m_i \in \mathbb{Z}, s \in \mathbb{N}\}$$

$$= \{\sum_{i=0}^{r-1} m_i\lambda^i, m_i \in \mathbb{Z}\} = \sum_{i=0}^{r-1} \mathbb{Z}\lambda^i \subseteq \mathbb{Z}[\lambda],$$

hence we get $\mathbb{Z}[\lambda] = \sum_{i=0}^{r-1} \mathbb{Z}\lambda^i$, i.e., $\mathbb{Z}[\lambda]$ is generated by $\{\lambda^i, \ 0 \leq i \leq r-1\}$ over $\mathbb{Z}$, as required.

$(2) \Rightarrow (1)$: Suppose $\lambda \in \mathbb{C}$ is such that $\lambda \neq 0$ and $\mathbb{Z}[\lambda]$ is generated by $\{x_i, 1 \leq i \leq m\}$ as a module over $\mathbb{Z}$. We can write $\lambda x_i = \sum_{j=1}^{m} a_{ji}x_j$ for some $a_{ij} \in \mathbb{Z}$, $1 \leq i \leq m$. This gives a system of linear equations over $\mathbb{Z}$, namely,

$$(\lambda - a_{11})x_1 - a_{21}x_2 - \cdots - a_{m1}x_m = 0$$

$$-a_{12}x_1 + (\lambda - a_{22})x_2 - \cdots - a_{m2}x_m = 0$$

$$\vdots$$

$$-a_{1m}x_1 - a_{2m}x_2 - \cdots + (\lambda - a_{mm})x_m = 0$$

having a non-trivial solution $(x_1, \cdots, x_m)$ in $\mathbb{C}$ and hence the coefficient matrix of the system must be singular, i.e., the determinant of the matrix

$$A = \begin{pmatrix} \lambda - a_{11} & -a_{12} & \cdots & -a_{1j} & \cdots & -a_{1m} \\ \vdots & \vdots & \vdots & & \vdots & \\ -a_{m1} & -a_{m2} & \cdots & -a_{mj} & \cdots & \lambda - a_{mm} \end{pmatrix}$$

must be zero which clearly implies that λ is a root of a monic polynomial with integer coefficients, as required. $\diamond$

3.8.7 Theorem: *The set of all algebraic integers in* $\mathbb{C}$ *is a subring of* $\mathbb{C}$, *called the* ring of algebraic integers.

Proof: Let λ and μ be algebraic integers, say, $\sum_{i=0}^{r} a_i \lambda^i = 0 = \sum_{j=0}^{s} b_j \mu^j$, for some a_i, $b_j \in \mathbb{Z}$ with $a_0 = 1 = b_0$. Now on repeated substitution for λ^r and μ^s, we have

$$
\mathbb{Z}[\lambda, \mu] = \{ \sum_{k=0}^{l} (\sum_{i+j=0}^{k} m_i n_j \lambda^i \mu^j \mid m_i, n_j \in \mathbb{Z} \text{ and } l \in \mathbb{N} \}
$$
$$
= \{ \sum_{i=0}^{r-1} \sum_{j=0}^{s-1} m_i n_j \lambda^i \mu^j \mid m_i, n_j \in \mathbb{Z} \} = \sum_{i=0}^{r-1} \sum_{j=0}^{s-1} \mathbb{Z} \lambda^i \mu^j,
$$

which means that $\mathbb{Z}[\lambda, \mu]$ is generated over $\mathbb{Z}$ by the finite set $\{\lambda^i \mu^j \mid 0 \le i \le r-1 \text{ and } 0 \le j \le s-1\}$ and hence $\mathbb{Z}[\lambda - \mu]$ and $\mathbb{Z}[\lambda\mu]$ which are $\mathbb{Z}$–submodules of $\mathbb{Z}[\lambda, \mu]$ are also finitely generated, consequently, $\lambda - \mu$ and $\lambda\mu$ are algebraic integers, as required. $\diamond$

Note: The ring of algebraic integers is *not* a field since 2 is an algebraic integer but $\frac{1}{2}$ is not.

3.8.8 Theorem: *Let G be a group of order n and (V, ρ) be a representation of G over* $\mathbb{C}$ *with its character $\chi = \chi_\rho$. Then*
1. *for each $x \in G$, $\chi(x)$ is an algebraic integer and*
2. *if V is irreducible with $d = \dim_{\mathbb{C}}(V)$ and h_x is the number of elements of the conjugacy class through $x \in G$, then $h_x \chi(x)/d$ is an algebraic integer.*

Proof: (1) By (3.7.5) above, we know that $\chi(x) = \mathrm{diag}(\zeta_1, \cdots, \zeta_d)$ with respect to a suitable basis of V where the ζ_i's are some m^{th}–roots of unity, $m = \mathrm{ord}(x)$. Hence $\chi(x) = \sum_{i=1}^{d} \zeta_i$ is a sum of algebraic integers and hence is an algebraic integer, as required.

(2) Let $C_1, \cdots, C_h$ be the conjugacy classes of G and $z_i = \sum_{x \in C_i} x$ so that $\{z_1, \cdots, z_h\}$ is a basis of the centre of $\mathbb{C}[G]$. It is obvious that $z_i z_j = \sum_{k=1}^{h} m_{ijk} z_k$, $m_{ijk} \in \mathbb{Z}^+$, where for each k, m_{ijk} is the number of pairs $(x, y) \in C_i \times C_j$ such that $xy \in C_k$. This implies that $A = \sum_{i=1}^{h} \mathbb{Z} z_i$ is a commutative subring of the centre of $\mathbb{C}[G]$ and is finitely generated as a $\mathbb{Z}$–module. Hence every element of A is integral over $\mathbb{Z}$, i.e., a root of some monic polynomial with integer coefficients.

Let $\bar{\rho} : \mathbb{C}[G] \to \text{End}_{\mathbb{C}}(V)$ be the *structure map* of the $\mathbb{C}[G]$–module V ((3.3.3) above). Take an $x \in C_i$. Since z_i is integral over $\mathbb{Z}$ and $\bar{\rho}$ is a homomorphism of $\mathbb{C}$–algebras, (in particular, $\mathbb{Z}$–linear), it follows that $\bar{\rho}(z_i)$ is integral over $\mathbb{Z}$. On the other hand, since z_i is in the centre of $\mathbb{C}[G]$, it follows that $\bar{\rho}(z_i)$ is $\mathbb{C}[G]$–linear, i.e., $\bar{\rho}(z_i) \in \text{End}_G(V)$. But $\text{End}_G(V) = \mathbb{C}$ by (3.3.11) above, since V is irreducible. Thus $\bar{\rho}(z_i) = \alpha_i \in \mathbb{C}$ is an algebraic integer and $\text{trace}(\bar{\rho}(z_i)) = d\alpha_i$. Now

$$
\begin{aligned}
d\alpha_i \;=\; & \text{trace}(\bar{\rho}(z_i)) \;=\; \sum_{x \in C_i} \text{trace}(\bar{\rho}(x)) \\
\;=\; & \sum_{x \in C_i} \text{trace}\rho(x) \;\; (\text{since } \bar{\rho}(y) = \rho(y),\; \forall\, y \in G) \\
\;=\; & \sum_{x \in C_i} \chi(x) \;=\; h_i \chi(x) \;\; (\text{since } \chi(y) = \chi(x),\; \forall\, y, x \in C_i).
\end{aligned}
$$

Thus $\alpha_x = h_x \chi(x)/d$ is an algebraic integer for every $x \in G$. ◇

3.8.9 Corollary: *The dimension of an irreducible representation of G over $\mathbb{C}$ divides the order of G.* (In fact, it divides the index of the centre of G. See (3.10.7) and and also Ex.(4.8.10) below).

Proof: Let V be an irreducible representation of G of dimension d with its character χ. Let $C_1, \cdots, C_h$ be the conjugacy classes of G with h_i the number of elements in C_i. Choose $x_i \in C_i, 1 \le i \le h$. Since V is irreducible, we have $\langle \chi, \chi \rangle = 1$. Thus we have

$$
\begin{aligned}
n \;=\; & n \langle \chi, \chi \rangle \;=\; n \frac{1}{n} \sum_{x \in G} \chi(x^{-1})\chi(x) \\
\;=\; & \sum_{i=1}^{h} (\sum_{x \in C_i} \chi(x^{-1})\chi(x)) \;=\; \sum_{i=1}^{h} \Big(h_i \chi(x_i^{-1})\chi(x_i) \Big).
\end{aligned}
$$

This gives that $n/d = \sum_{i=1}^{h} \chi(x_i^{-1})(h_i \chi(x_i)/d)$ is an algebraic integer since it is a sum of products of algebraic integers. But n/d is a rational number and so by (3.8.5) above, $n/d \in \mathbb{Z}$, i.e., $d \mid n$, as required. ◇

3.8.10 Theorem: *Let (V, ρ) be an irreducible representation of dimension d with its character χ. Let $x \in G$ be such that the conjugacy class C_x of x has h_x elements with $\gcd(h_x, d) = 1$. Then either $\chi(x) = 0$ or $\rho(x)$ is a scalar multiplication on V.*

Proof: By (3.4.2) above, choose a suitable basis for V with respect to which the matrix of $\rho(x) = \text{diag}(\zeta_1, \cdots, \zeta_d)$ where the ζ_i's are some m^{th}-roots of 1, $m = \text{ord}(x)$. Let ζ be a primitive m^{th}-root of 1. Then we have $\zeta_i = \zeta^{m_i}$ for some $m_i \in \mathbb{Z}$ with $0 \leq m_i \leq m - 1$, $1 \leq i \leq d$.

If all the ζ_i's are equal, $\rho(x)$ is a scalar matrix and we are through. Assume that not all the ζ_i's are equal. We shall prove that $\chi(x) = 0$. Since $\gcd(h_x, d) = 1$, there exist integers a and b such that $ah_x + bd = 1$ and hence

$$\frac{\chi(x)}{d} = \frac{ah_x \chi(x)}{d} + b\chi(x) = \alpha \text{ (say)}$$

is an algebraic integer by (3.8.8) above. Since the ζ^{m_i}'s are all of modulus 1, we get that $\mid \zeta^{m_1} + \cdots + \zeta^{m_d} \mid \; \leq d$ with equality if and only if $m_1 = \cdots = m_d$, i.e., $\zeta_1 = \cdots = \zeta_d$. But the ζ_i's are not all equal and hence we get that

$$\mid \alpha \mid = \mid \frac{\chi(x)}{d} \mid = \mid \frac{\sum_{i=1}^{d} \zeta_i}{d} \mid = \mid \frac{\sum_{i=1}^{d} \zeta^{m_i}}{d} \mid \; < 1.$$

If $f(X) = X^r + a_1 X^{r-1} + \cdots + a_{r-1} X + a_r$ with $a_i \in \mathbb{Z}$ is the minimal polynomial of α, (i.e., monic and of least degree satisfied by α), then it can be seen that $f(X)$ is irreducible (as an immediate consequence of *Gauss' Lemma* (Ex.(3.11.21) below)) and any root β of $f(X)$ in $\mathbb{C}$ which is obviously an algebraic integer is of the form $\beta = \varphi(\alpha)$ for some automorphism φ of the field $\mathbb{Q}[\zeta]$ (Ex.(3.11.22) below). But then we have $\varphi(\zeta) = \zeta^j$ for some j coprime to m. Now we have $\beta = \varphi(\alpha) = (\sum_{i=1}^{d} \varphi(\zeta^{m_i}))/d = (\sum_{i=1}^{d} \zeta^{jm_i})/d$. But then it follows that $\mid \beta \mid = \mid (\sum_{i=1}^{d} \zeta^{jm_i})/d \mid \; < 1$ since the ζ^{jm_i}'s are all of modulus 1 and not all equal. Hence we get that the product of the roots of $f(X)$ is $(-1)^r a_r$ which is an integer of modulus strictly less than 1 since each root of $f(X)$ is of modulus less than 1. This is not possible (by (3.8.5) above) unless $a_r = 0$, but then by the irreducibility of $f(X)$, it means that $r = 1$, i.e., $f(X) = X$ or $\alpha = \chi(x)/d = 0$. ∎

3.9   Burnside's $p^a q^b$-Theorem

As an application of the theory of complex characters, we shall prove a well-known theorem of Burnside on the solvability of groups G whose orders have at the most two distinct prime divisors, i.e., $|G| = p^a q^b$

with p and q primes and $a, b \in \mathbb{Z}^+$. This is also known as the $p^a q^b$-Theorem of Burnside.

3.9.1 Solvable groups: A group G is said to be *solvable* if there exists a finite descending sequence of subgroups starting with G and ending with $\{1\}$, say $(\star)$: $G = G_0 \supset G_1 \supset \cdots \supset G_r = \{1\}$ such that for each i, G_i is normal in G_{i-1} and the quotient G_{i-1}/G_i is abelian, $1 \leq i \leq r$. Such a sequence $(\star)$ is called a *solvable series* for G.

3.9.2 Example: Any abelian group is solvable.

3.9.3 Remark: A simple group is solvable if and only if it is cyclic of prime order, consequently, the *alternating group A_n* is not solvable for $n \geq 5$ (it being simple and non–abelian).

3.9.4 Theorem: (i) *Subgroups and quotient groups of a solvable group are solvable.*
(ii) *If a group G has a normal subgroup H such that both H and G/H are solvable, then G is solvable.*

Proof: Exactly similar to the case of groups or modules of finite length (1.4.7) above. ◇

3.9.5 Corollary: *Any p-group is solvable since its centre is non-trivial, abelian and normal such that the quotient by the centre is a p-group of lower order.*

3.9.6 Theorem (Burnside): *Any group of order $p^r q^s$ is solvable for any primes p and q.*

Proof: Let G be of order $n = p^r q^s$. If $p = q$, then G is a p-group and so we are through. We may assume that $p \neq q$ and $r, s \geq 1$. Since any subgroup or a quotient of G is of order of the same type as that of G, it suffices to show that G has a non–trivial normal subgroup.

Take any *Sylow p-subgroup H* of G. Since H is a p-group (of order p^r), its centre is non–trivial. Pick up an element $x \neq 1$ in the centre of H. We distinguish two possibilities.

Case 1 : Suppose x *is in the centre of G.*

Let K be the subgroup generated by x. Then K is obviously a non–trivial normal subgroup of G that we are looking for.

Case 2 : Suppose x *is not in the centre of* G.

Note then that G is not abelian and so its *commutator subgroup* $G^{(1)} \neq \{1\}$. Further, the *centraliser* $C_G(x)$ of x is a proper subgroup of G containing H and hence has index q^t for some $t, 1 \leq t \leq s$. This means that the conjugacy class C_x of x has $h_x = q^t$ elements.

Let $\chi_1, \cdots, \chi_h$ be all the irreducible characters of G over $\mathbb{C}$, i.e., the irreducible components of the regular representation of G, with respective dimensions $1 = m_1, \cdots, m_h$, χ_1 being the character of the trivial representation, where h is the number of conjugacy classes of G. We have $\chi = \chi_{\text{reg}} = \sum_{i=1}^{h} m_i \chi_i$. Since $x \neq 1$, recall (by (3.4.3)(2) above) that we have

$$0 = \chi(x) = \sum_{i=1}^{h} m_i\, \chi_i(x) = 1 + \sum_{i=2}^{h} m_i\, \chi_i\,(x) \Rightarrow \chi_i(x) \neq 0$$

for some $i \geq 2$. Note that if $q \mid m_i$ for every $i \geq 2$ such that $\chi_i(x) \neq 0$, then it follows that $0 = 1 + q\alpha$ where $\alpha = \sum_{i=2}^{h}(m_i/q)\chi_i(x)$ which is an algebraic integer by (3.8.8) above. But this is not possible by (3.8.5) above, since $\alpha = -1/q \in \mathbb{Q}$ is not an integer. Thus we can choose an $i \geq 2$ such that $\chi_i(x) \neq 0$ and q not dividing m_i.

Let $(V, \rho) = (V_i, \rho_i)$ be the representation space of the irreducible character χ_i. By our choice of i, we have $\gcd(q, m_i) = \gcd(h_x, m_i) = 1$ (since $h_x = q^t$). Since $\chi_i(x) \neq 0$, we get by (3.8.10) above, that $\rho(x)$ is a scalar multiplication on V. Since ρ is not the trivial representation, we have $\text{Ker}(\rho) \neq G$.

If $m_i = 1$, we have $\{1\} \neq G^{(1)} \subseteq \text{Ker}\rho \neq G$ and so we can take $K = G^{(1)}$ which is a non–trivial normal subgroup of G. On the other hand, if $m_i \geq 2$, we know that $\rho(G)$ cannot consist of only scalar multiplications on V (otherwise, any vector subspace of V would be a G–subspace, contradicting the irreducibility of V). Now look at the subgroup of G, namely, $K = \{y \in G \mid \rho(y) = \ell_{\lambda(y)} \text{ on } V, \lambda(y) \in \mathbb{C}\}$, which is obviously a normal subgroup of G such that $K \neq G$ and $x \in K$, i.e., K is a non–trivial normal subgroup of G, as required. $\Diamond$

3.9.7 Remark: In 1963, Walter Feit and John Thompson proved a remarkable theorem (conjectured by Burnside early in this century) that "GROUPS OF ODD ORDER ARE SOLVABLE".

(i) This is the title of their paper that occupied a complete issue of the Pacific Journal of Mathematics, Vol. 13 (1963), pp. 775–1029. The proof runs into 255 pages, an odd number again. It is referred to as the odd order paper!

(ii) The theorem proved above is subsumed by this paper only for primes p and q both odd. ∎

3.10 Tensor Product of Representations

We conclude this chapter with a collection of some of the basic facts about the *tensor product of representations*. The proofs of most of them are rather easy and hence omitted. All groups considered are supposed to be finite (though some of the statements are valid for other groups as well).

3.10.1 Group algebra of $G \times H$: *The group algebra of the direct product of groups is naturally isomorphic to the tensor product of their group algebras (1.10.17), i.e, we have a K–algebra isomorphism,*

$$K[G \times H] \longrightarrow K[G] \otimes_K K[H], \left(\sum a_{gh}(g,h)\right) \mapsto \sum a_{gh}g \otimes h.$$

The given map is obviously a K–algebra homomorphism and is a bijection of the K–basis $G \times H$ of $K[G \times H]$ onto the K–basis $\{g \otimes h \mid (g,h) \in G \times H\}$ of $K[G] \otimes_K K[H]$ (1.10.15)(3), as required.

• By (3.1.7) above, it is obvious that $K[G \times H]$ *is semi–simple if and only if both $K[G]$ and $K[H]$ are so.*

3.10.2 Conjugacy classes of $G \times H$: *The conjugacy class through an element $(g,h) \in G \times H$ is given by $C_{(g,h)} = C_g \times C_h$ where C_g is the conjugacy class through g in G, etc. In particular, the number of conjugacy classes of $G \times H$ is equal to the product of the number of conjugacy classes of G and H, i.e., $h(G \times H) = h(G)h(H)$.*

We assume for the rest of this section that K is algebraically closed *and that $K[G \times H]$ is semi-simple.*

3.10.3 Product representations of $G \times H$: Given representations (V, θ) and (W, ϑ) of G and H respectively (over the same field K), we have a natural representation $(V \otimes W, \theta \otimes \vartheta)$ of $G \times H$, defined by $\big((\theta \otimes \vartheta)(g, h)\big)(v \otimes w) = \theta(g)(v) \otimes \vartheta(h)(w)$ for all $(g, h) \in G \times H$ and $(v, w) \in V \times W$. This is called the *product representation* of $G \times H$, *afforded by* (V, θ) *and* (W, ϑ).

Note that an arbitrary representation of $G \times H$ need *not* be of product type. (Give an example.)

3.10.4 Characters of product representations: By Ex.(2.9.12) above, *the character of the product representation* $(V \otimes W, \theta \otimes \vartheta)$ *of $G \times H$ is given by* $\chi_{\theta \otimes \vartheta} = \chi_\theta \cdot \chi_\vartheta$, *i.e.,* $\chi_{\theta \otimes \vartheta}(g, h) = \chi_\theta(g)\chi_\vartheta(h)$ *for all* $(g, h) \in G \times H$. *In particular, product of trivial representations is trivial and that of characters (i.e., one-dimensional representations) is a character.*

3.10.5 *The inner product* (3.6.4) *of the characters of the product representations is the product of the respective inner products, i.e.,*

$$\Big\langle \chi_{\theta_1 \otimes \vartheta_1}, \ \chi_{\theta_2 \otimes \vartheta_2} \Big\rangle_{G \times H} = \Big\langle \chi_{\theta_1}, \ \chi_{\theta_2} \Big\rangle_G \Big\langle \chi_{\vartheta_1}, \ \chi_{\vartheta_2} \Big\rangle_H$$

where (V_i, θ_i) and (W_i, ϑ_i) are representations of G and H respectively, $i = 1, 2$.

3.10.6 Product of irreducible representations: *Product of* ordinary *irreducible representations is irreducible (by* (3.7.2) *and* (3.10.5)). *Consequently, if*

$$\mathrm{Irr}_K(G) = \{(V_i, \theta_i)\}_{i=1}^{h(G)} \quad \text{and} \quad \mathrm{Irr}_K(H) = \{(W_j, \vartheta_j)\}_{j=1}^{h(H)}$$

are complete sets of mutually inequivalent irreducible ordinary representations of G and H respectively, then (for the group $G \times H$),

$$\mathrm{Irr}_K(G \times H) = \{(V_i \otimes W_j, \theta_i \otimes \vartheta_j) \mid 1 \leq i \leq h(G) \text{ and } 1 \leq j \leq h(H)\}.$$

Note that this is *not* true if Char $K \neq 0$. (Give an example.)

3.10.7 Theorem: *The dimension of an ordinary irreducible representation of a group G divides the index of the centre of G.* (This generalises (3.8.9) above. See also Ex.(4.8.10) below, for a further generalisation.)

Proof(J.Tate): Let (V, θ) be an ordinary irreducible representation of G. Let $Z = Z(G)$ be the centre of G. By (3.3.11)(2) above, for each $x \in Z$, $\theta(x)$ is a scalar multiplication on V by some $\alpha(x) \in K$. But then $\alpha(x)$ is a root of unity. Furthermore, $\alpha \in \widetilde{Z}$.

For each positive integer m, consider the product representation $(V_m, \theta_m) = (V^{\otimes m}, \theta^{\otimes m})$ of G^m which is irreducible (by (3.10.6) above). It is clear that for all $x_j \in Z$ and $w \in V_m$, we have

$$\theta_m(x_1, \cdots, x_m)(w) = (\alpha(x_1) \cdots \alpha(x_m))w = \alpha(x_1 \cdots x_m)w.$$

Let $H_m = \{(x_1, \cdots, x_m) \in Z^m \mid x_1 \cdots x_m = 1\}$ which is a subgroup of Z^m, isomorphic to Z^{m-1}. Since $H_m \subseteq \mathrm{Ker}(\theta_m)$, θ_m goes down to a representation of G^m/H_m on V_m which is obviously irreducible.

By Ex.(3.11.24) below, we get that (3.8.9) above is true (for K in place of $\mathbb{C}$) for the irreducible representation V_m of G^m/H_m, i.e., d^m divides n^m/r^{m-1} where $n = |G|$, $r = |Z|$ and $d = \dim_K(V)$. This gives that $(n/rd)^m \in \mathbb{Z} \cdot (1/r)$ for all m, i.e., the ring $\mathbb{Z}[n/rd]$ is finitely generated as a $\mathbb{Z}$–module, being a subgroup of a cyclic group $\mathbb{Z} \cdot (1/r)$. But then by (3.8.6) and (3.8.5) above, n/rd is an integer. $\Diamond$

3.10.8 Tensor product of representations of G: Given two representations (V_1, θ_1) and (V_2, θ_2) of G, we have a natural representation $(V_1 \otimes V_2, \theta_1 \otimes \theta_2)$ of G, defined by

$$\Big((\theta_1 \otimes \theta_2)(g)\Big)(v_1 \otimes v_2) = \theta_1(g)(v_1) \otimes \theta_2(g)(v_2)$$

for all $g \in G$ and $(v_1, v_2) \in V_1 \times V_2$. This is called the *tensor product representation* of G, afforded by the representations (V_i, θ_i), $i = 1, 2$.

Note that the tensor product of two representations of G is simply the *restriction* of the corresponding product representation of $G \times G$ (3.10.3) to the *diagonal subgroup* $\Delta_G = \{(g, g) \in G \times G \mid g \in G\}$ of

$G \times G$. For this reason, it is also called the tensor product representation for the *diagonal action* of G.

3.10.9 Remarks: 1. As in (3.10.4) and by restriction, *the character of the tensor product of two representations of G is the product of the respective characters.*

2. However, *unlike* in (3.10.5), the inner product of the characters of two tensor product representations is *not* equal to the product of the inner products. (This is not surprising since there is no natural way to break the inner product into factors.)

3. *Most important*: unlike in (3.10.6), *even for the case of* *ordinary* *representations of G, tensor product of irreducible representations is usually not irreducible except in trivial special cases like when one of them is a character* (Ex.(3.11.15) below).

4. *Decomposing the tensor product $V \otimes_K W$ of* even ordinary *irreducible representations $V, W \in \mathrm{Irr}_K(G)$ into irreducible components, say,* $(\star): \ V \otimes_K W = \bigoplus_{S \in \mathrm{Irr}_K(G)} S^{\oplus(n_S)}$, *i.e., finding the decomposition numbers n_S,* (3.3.10)(5)), *is a difficult problem and there are no general methods available, yet it is one of the most useful information one looks for. The decomposition $(\star)$ above, is called the* Clebsch-Gordan series *for $V \otimes_K W$.*

In this book, we have not considered the decomposition problem at all even in the special cases of the groups like S_n, etc. Nevertheless, just for the case of the group S_n itself, this problem is solved completely and the solution is given by the so called "Littlewood–Richardson Rule" (cf. [25])).

3.10.10   Contragredient representation: Given a representation (V, ϱ) of G over K, the representation $(V^\star, \varrho^\star)$ on the *dual $V^\star$* $= \mathrm{Hom}_K(V, K)$, defined by $(\varrho^\star(g)f)(v) = (gf)(v) = f(g^{-1}v)$ for all $g \in G$, $f \in V^\star$ and $v \in V$, is called the *dual* or the *contragredient* representation or the *representation contragredient* to V.

It is trivial to see that if $\varrho(g) = \left(\varrho_{ij}(g)\right)$ is the matrix representation with respect to some chosen basis $\{v_i\}$ of V, then $\varrho^\star(g) = {}^t(\varrho(g^{-1}))$ is the transpose inverse of $\left(\varrho_{ij}(g)\right)$ with respect to the *dual basis* $\{v_i^\star\}$ of $V^\star$.

3.10.11 Remarks: 1. Given representations V and W of G, we have a natural representation of G on $\text{Hom}_K(V, W)$, defined by $(gf)(v) = gf(g^{-1}v)$, $\forall g \in G$, $f \in \text{Hom}_K(V, W)$ and $v \in V$. The contragredient representation V^* is a special case of this for the choice of the trivial representation $W = K$.

2. The *character* χ_{V^*} of the contragredient representation V^* is given by $\chi_{V^*}(g) = \chi_V(g^{-1})$, $\forall g \in G$ (Ex.(3.11.19) below).

3. For any irreducible representation V of G, $V \otimes_K V^*$ is isomorphic to an isotypical component of $K[G]$ and conversely. In other words, $\oplus_{V \in \text{Irr}_K(G)} V \otimes_K V^*$ is the isotypical decomposition of $K[G]$.   ∎

3.11   Exercises

Unless otherwise stated explicitly, a ring R means a ring with 1 (commutative or not). All representations considered are supposed to be over a fixed ground field K (of arbitrary characteristic) and finite dimensional too.

1. Let $R[G]$ be the group algebra of a group G (finite or not), over a commutative ring R. For $z = (\sum_{x \in G} a_x x) \in R[G]$, recall that $\text{supp}(z) = \{x \in G \mid a_x \neq 0\}$ (which is a finite set) is called the *support* of z. Show that (i) $\text{supp}(gz) = g(\text{supp}(z))$ for all $g \in G$ and $z \in R[G]$. (ii) $x \in (\text{supp}(z)) \iff 1 \in (\text{supp}(x^{-1}z)) \iff 1 \in (\text{supp}(zx^{-1}))$. (iii) If z is central in $R[G]$, then the subgroup generated by $\text{supp}(z)$ is normal in G.

2. Show that the group algebra $R[G]$ is naturally isomorphic to the ring S of all R-valued maps of *finite support* defined on G, i.e., $S = \{f : G \to R \mid |\text{supp}(f)| < \infty\}$ (where $\text{supp}(f) = \{x \in G \mid f(x) \neq 0\}$ is called the *support* of f). (The natural map $x \mapsto \psi_x$, where $\psi_x(y) = 1$ or 0 according as $y = x$ or $y \neq x$, is an isomorphism.)

3. Show that the map $(\sum_{x \in G} a_x x) \mapsto (\sum_{x \in G} a_x x^{-1})$ of $R[G]$ onto its opposite $(R[G])^0$ (1.1.1) is an *anti-isomorphism* of rings.

4. Show that $R[G]$ is left Artinian (resp. Noetherian) if and only if it is right Artinian (resp. Noetherian).

    Remark: A non-trivial theorem of Connell (cf. [42]) says that if K is a field, then $K[G]$ is Artinian if and only if G is finite. (First the difficult part, show the existence of a *finite* normal subgroup H of G

and then proceed by induction on the length of the ring $K[G]$ ((2.4.9) above), use the surjectivity of the natural map $K[G] \to K[G/H]$, etc.)

5. Given a subgroup H of G, let $\pi_H \colon R[G] \to R[H]$, $z = (\sum_{x \in G} a_x x) \mapsto (\sum_{y \in H} a_y y) = \pi_H(z)$. Show the following.

   (i) π_H is an R–linear projection onto $R[H]$ but not necessarily a homomorphism of rings.

   (ii) $H \cap (\mathrm{supp}(z')) = \emptyset$ where $z' = z - \pi_H(z)$.

   (iii)   If H is normal in G, then $\pi_H(xzx^{-1}) = x\pi_H(z)x^{-1}$ for all $z \in R[G]$ and $x \in G$.

   (iv) $z \in R[H]$ is a unit in $R[H]$ $\iff$ z is a unit in $R[G]$.

   (v) $z \in R[H]$ is a zero–divisor in $R[H]$ $\iff$ z is so in $R[G]$.

   (Hints: (i) $\pi_H(tz) = t\pi_H(z)$, $\forall\, t \in H$, $z \in R[G]$. (ii) $z \in R[H]$, $zz' = 0$ in $R[G]$, $z' \neq 0$, choose $x \in G$ so that $1 \in (\mathrm{supp}(z'x))$. Note that $z\pi_H(z'x) = 0$ but $\pi_H(z'x) \neq 0$.)

6. Show that $M = \{\sum_{x \in G} a_x x \in R[G] \mid \sum_{x \in G} a_x = 0\}$ is a 2–sided ideal of $R[G]$, called the *augmentation ideal* of $R[G]$. Is this ideal a *direct summand* in $R[G]$?

   Suppose that K is a field. Then show that M is of codimension 1 in $K[G]$ and hence it is *maximal* as a left/right/2–sided ideal. Give an example of a maximal left ideal which is *not* of codimension 1.

7. Give an example of an *infinite group* G such that $K[G]$ is *not* semi–simple even if K is algebraically closed and of characteristic 0. (Hint: Take $G = \mathbf{R}^{>0}$, the multiplicative group of positive real numbers and $K = \mathbf{C}$. The $\mathbf{C}[G]$–module $\mathbf{C}^2$, under the scalar multiplication,

$$\left(\sum_{x \in G} a_x x\right)(\alpha, \beta) = \left(\alpha + \left(\sum_{x \in G} a_x \log x\right)\beta, \beta\right)$$

has the submodule $\mathbf{C} \times \{0\}$ which is not a direct summand.)

Henceforward, G denotes a finite group and K a field.

8. Let H be a subgroup of G and $a_H = \sum_{h \in H} h \in K[G]$. Then show that $\dim_K(K[G]a_H) = [G : H]$, the index of H in G.

9. For $z \in K[G]$, let $\mathrm{trace}(z)$ be the *trace* of the K–linear endomorphism ℓ_z (left multiplication by z) of $K[G]$. Show that $\mathrm{trace}(\alpha\beta) = \sum_{x \in G} a_x b_{x^{-1}}$ for all $\alpha = \sum_{x \in G} a_x x$ and $\beta = \sum_{x \in G} b_x x$ in $K[G]$. Hence or otherwise deduce the following.

   (1) $\mathrm{trace}(\alpha) = 0$ if α is *nilpotent* in $K[G]$ (but not conversely),

(ii) trace(e) lies in the *prime subfield* of K for an *idempotent* $e \in K[G]$ and

(iii) trace(e) = $\dim_K(K[G]e)/|G|$ for an idempotent $e \in K[G]$ where $|G|$ is the order of the group G.

10. Recall that a group is said to be *dihedral* (of order 2ℓ) and denoted by $\mathbf{D}_{2\ell}$ if it is generated by two elements a and b with $a^\ell = 1$, $b^2 = 1$ and $a^{-1} = bab$. Prove the following.
    (i) $\mathbf{D}_{2\ell} = \{a^i b^j \mid 0 \leq i \leq \ell - 1 \text{ and } 0 \leq j \leq 1\}$.
    (ii) $\mathbf{D}_{2\ell}$ is abelian $\iff$ b is central $\iff$ $\ell = 2$, in which case $\mathbf{D}_4 \approx \mathbf{Z}/2\mathbf{Z} \times \mathbf{Z}/2\mathbf{Z}$.
    (iii) If $\ell = 3$, show that $\mathbf{D}_6 \approx S_3$ where S_3 is the *permutation group* on 3 symbols.
    (iv) If $\ell = 4$, show that $\mathbf{D}_8$ is *not* isomorphic to the *Quaternion group* $Q_8 = \{\pm 1, \pm i, \pm j, \pm k\} \subset \mathbf{H} = \mathbf{R}[i, j, k]$ (Ex.(2.9.12) above).

11. Determine the conjugacy classes of $\mathbf{D}_4$, S_3, $\mathbf{D}_8$ and Q_8.

12. Let K be a field of characteristic $\neq 2$. Prove the following.
    (i) The map $K[\mathbf{D}_4] \to K^4$, sending $a \mapsto (1, -1, 1, -1)$ and $b \mapsto (1, 1, -1, -1)$, induces an isomorphism of K-algebras.
    (ii) Let $\ell \geq 3$ and assume that K contains ℓ^{th} roots of unity, say ζ_ℓ is a *primitive* ℓ^{th} root. Then the map $K[\mathbf{D}_6] \to K^2 \times M_2(K)$, sending $a \mapsto (1, 1, \operatorname{diag}(\zeta_3, \zeta_3^{-1}))$ and $b \mapsto (1, -1, \operatorname{antidiag}(1, 1))$, induces an isomorphism of K-algebras.
    (iii) Describe an isomorphism between $K[\mathbf{D}_8]$ and $K^4 \times M_2(K)$, as in the case of $\mathbf{D}_6$ outlined above.
    (iv) Let K be a subfield of $\mathbf{R}$ and $\mathbf{H}_K = K[i, j, k] \subseteq \mathbf{R}[i, j, k]$. Then $K[Q_8] \approx K^4 \times \mathbf{H}_K$. Describe an isomorphism. Choosing $K = \mathbf{R}$ and using the fact that $\mathbf{C}$ is a *splitting field* for $\mathbf{H} = \mathbf{H}_\mathbf{R}$ (2.8.12), deduce that $\mathbf{C}[Q_8] \approx \mathbf{C}^4 \times M_2(\mathbf{C})$.
    (v) $\mathbf{C}[Q_8] \approx \mathbf{C}[\mathbf{D}_8]$ though the groups themselves are *not* isomorphic.

13. Show that $\mathbf{C}[G] \approx \mathbf{C}^n$ for all abelian groups G of order n.

These examples show, in particular, that the group G cannot be recovered from the group algebra $K[G]$. However, the following is of special interest.

14. For *finite abelian* groups G and G', show that $\mathbf{Z}[G] \approx \mathbf{Z}[G']$ $\iff$ $G \approx G'$. (Use the (non-trivial) fact that if a unit $u \in \mathcal{U}(\mathbf{Z}[G])$ is of *finite order*, then $\pm u \in G$ (cf. [42]).)

15. Let $\widetilde{G}$ be the *character group* (3.5.6) of G. Show that the commutator subgroup $G^{(1)}$ of G is given by $G^{(1)} = \bigcap_{\chi \in \widetilde{G}} \mathrm{Ker}(\chi)$. Deduce that $G^{(1)}$ and $\widetilde{G}$ are determined by each other.

16. Let $\rho \in \widetilde{G}$ and (V, ϑ) be a representation of G. Show that ϑ is irreducible if and only if $\vartheta \otimes \rho$ is irreducible.

17. Given a representation (V, ϑ) and a character $\rho \in \widetilde{G}$, show that the *inner products* (3.6.4) $\left\langle \chi_\vartheta, \chi_\vartheta \right\rangle$ and $\left\langle \chi_{\vartheta \otimes \rho}, \chi_{\vartheta \otimes \rho} \right\rangle$ are equal.

18. Let G act on a set X and χ_X be the character of the *permutation representation* of G over $K = \mathbb{C}$ afforded by X (3.3.5)(3). Prove the following. (i) $\chi_X(g) = |\{x \in X \mid gx = x\}|$, $\forall\, g \in G$.
(ii) If χ_1 is the *trivial* character of G, then $\left\langle \chi_X, \chi_1 \right\rangle$ is the number of G-orbits in X; in particular, G acts *transitively* on X if and only if $\left\langle \chi_X, \chi_1 \right\rangle = 1$.
(iii) If G acts transitively on X and Y, consider the *diagonal action* of G on $X \times Y$, denoted by $X \times_G Y$. Then show that $\left\langle \chi_X, \chi_Y \right\rangle = \left\langle \chi_{X \times_G Y}, \chi_1 \right\rangle$ (which is the number of G-orbits in $X \times_G Y$).

19. Let V^* be the contragredient dual of a representation V of G (3.10.10). Prove the following. (i) $\chi_{V^*}(x) = \chi_V(x^{-1})$, $\forall\, x \in G$.
(ii) There is a natural bijective correspondence between G-quotients of V and G-submodules of V^*; in particular, V is irreducible if and only if V^* is so.
(iii) Show that $\left\langle \chi_V, \chi_V \right\rangle = \left\langle \chi_{V^*}, \chi_{V^*} \right\rangle$ (which gives another proof that V is irreducible if and only if V^* is so in case Char $K = 0$).

20. Given representations V and W of G, let $\mathrm{Hom}_K(V, W)$ be the representation as in (3.10.11), i.e., $(gf)(v) = gf(g^{-1}v)$, for all $g \in G$, $f \in \mathrm{Hom}_K(V, W)$ and $v \in V$. Prove the following.
(i) The space of G-linear homomorphisms from V to W is the same as the space of G-invariants in $\mathrm{Hom}_K(V, W)$, i.e., $\mathrm{Hom}_G(V, W) = (\mathrm{Hom}_K(V, W))^G$.
(ii) The *multiplicity* with which V occurs in W is equal to the dimension of $\mathrm{Hom}_G(V, W)$ if V is *irreducible*.
(iii) The natural K-linear isomorphism $V^* \otimes_K W \xrightarrow{\sim} \mathrm{Hom}_K(V, W)$, as given in Ex.(1.11.55) above, is a $K[G]$-linear isomorphism, i.e., $\mathrm{Hom}_K(V, W) \cong V^* \otimes_K W$ and hence $\chi_{\mathrm{Hom}_K(V,W)} = \chi_{V^*} \chi_W$.

21. **Gauss' Lemma:** Let $f(X) = a_0 X^n + \cdots + a_{n-1} X + a_n \in \mathbb{Z}[X]$ be a non–constant polynomial such that $\gcd(a_0, \cdots, a_n) = 1$. Then $f(X)$ is irreducible in $\mathbb{Z}[X]$ if and only if it is irreducible in $\mathbb{Q}[X]$.

22. A complex number which is a root of a non–constant polynomial with rational coefficients is called an *algebraic number*. Prove the following.
    (i) $\alpha \in \mathbb{C}$ is an algebraic number $\Longleftrightarrow$ the subring $\mathbb{Q}[\alpha]$ of $\mathbb{C}$ is finite dimensional over $\mathbb{Q}$ $\Longleftrightarrow$ $\mathbb{Q}[\alpha]$ is a field.
    (ii) Some integral multiple of any algebraic number is an algebraic integer.
    (iii) The set of all algebraic numbers is the field of fractions of the ring of all algebraic integers and it is also the algebraic closure $\overline{\mathbb{Q}}$ of $\mathbb{Q}$.
    (iv) Two algebraic numbers α and β are roots of the same irreducible polynomial $f(X) \in \mathbb{Q}[X]$ if and only if there exists an isomorphism of the fields, $\sigma \colon \mathbb{Q}[\alpha] \to \mathbb{Q}[\beta]$ such that $\sigma(\alpha) = \beta$. (Is the irreducibility of $f(X)$ essential?)

23. Show that a group is *not* simple if the number of elements in some conjugacy class is a positive power of a prime. (Imitate Case 2 in the proof of (3.9.6) above.) Is the converse true for a non–abelian non–simple group?

24. Let K be an algebraically closed field K of characteristic 0. Show that $\overline{\mathbb{Q}} \subseteq K$ and that all the results of §3.8 are valid for K in place of $\mathbb{C}$.

25. Let $G = <x>$ be the cyclic group of order ℓ, generated by x. Let $\{\zeta_i\}_{i=1}^\ell$ be the set of all the ℓth root of 1 in $\mathbb{C}$ with $\zeta_1 = 1$. Show that the character table of $<x>$ is the following.

	1	x	x^2	$\cdots$	$x^{\ell-1}$
χ_1	1	1	1	$\cdots$	1
χ_2	1	ζ_2	ζ_2^2	$\cdots$	$\zeta_2^{\ell-1}$
$\vdots$	$\vdots$	$\vdots$	$\vdots$	$\vdots$	$\vdots$
χ_j	1	ζ_j	ζ_j^2	$\cdots$	$\zeta_j^{\ell-1}$
$\vdots$	$\vdots$	$\vdots$	$\vdots$	$\vdots$	$\vdots$
χ_ℓ	1	ζ_ℓ	ζ_ℓ^2	$\cdots$	$\zeta_\ell^{\ell-1}$

26. Find the complex character tables of $\mathbf{D_4}$ and $\mathbf{D_6} = S_3$.

27. Find the complex character table of the Quaternion group Q_8.

28. Let $G = \mathbf{D}_{2\ell}$. Prove the following.
    (i)   The commutator subgroup $\mathbf{D}_{2\ell}^{(1)}$ is given by

$$\mathbf{D}_{2\ell}^{(1)} = \begin{cases} <a> & \text{if } \ell \text{ is odd,} \\ <a^2> & \text{if } \ell \text{ is even.} \end{cases}$$

Consequently, $\mathbf{D}_{2\ell}$ has exactly 2 or 4 group characters according as ℓ is odd or even.
(ii) The number of conjugacy classes is given by

$$h(\mathbf{D}_{2\ell}) = \begin{cases} (\ell+3)/2 & \text{if } \ell \text{ is odd,} \\ (\ell+6)/2 & \text{if } \ell \text{ is even.} \end{cases}$$

(iii) The dimensions of the irreducible representations are all ≤ 2 for every ℓ.

29. Let $\ell = 2m$ be *even*. Show that the conjugacy classes of $\mathbf{D}_{2\ell}$ are given by $C_0 = \{1\}$, $C_1 = \{a, a^{-1}\}$, $C_2 = \{a^2, a^{-2}\}$, $\cdots$, $C_{m-1} = \{a^{m-1}, a^{-m+1}\}$, $C_m = \{a^m\}$, $C_b = \{a^{2k}b \mid 0 \leq k \leq m-1\}$ and $C_{ab} = \{a^{2k+1}b \mid 0 \leq k \leq m-1\}$.
    • Show that the character table of $\mathbf{D}_8$ is given as follows.

	C_0	C_1	C_2	C_b	C_{ab}
χ_1	1	1	1	1	1
χ_2	1	1	1	-1	-1
χ_3	1	-1	1	1	-1
χ_4	1	-1	1	-1	1
χ_5	2	0	-2	0	0

Verify that the tables are the same for the non–isomorphic groups $\mathbf{Q}_8$ and $\mathbf{D}_8$ (Exs.(3.11.10) and (3.11.12) above).

30. Let $\ell = 2m+1$ be *odd*. Show that the conjugacy classes of $\mathbf{D}_{2\ell}$ are given by $C_0 = \{1\}$, $C_1 = \{a, a^{-1}\}$, $C_2 = \{a^2, a^{-2}\}$, $\cdots$, $C_m = \{a^m, a^{-m}\}$ and $C_b = \{a^i b \mid 0 \leq i \leq 2m\}$.
    Furthermore, show that the character table of $\mathbf{D}_{10}$ is given as follows (wherein $\zeta \neq 1$ is a 5th root of 1).

	C_0	C_1	C_2	C_b
χ_1	1	1	1	1
χ_2	1	1	1	-1
χ_3	2	$\zeta+\zeta^{-1}$	$\zeta^2+\zeta^{-2}$	0
χ_4	2	$\zeta^2+\zeta^{-2}$	$\zeta+\zeta^{-1}$	0

3.12 True/False Statements

Determine which of the following statements are true (T) or false (F) or partially true (PT). Justify your answers by giving a proof if (T) or providing a counter–example if (F)/(PT) or supplying the additional hypothesis needed to make it (T) (along with a proof) if (PT), as the case may be.

In what follows, G stands for a finite group and K an algebraically closed field such that K[G] is semi–simple. All representations considered are assumed to be over K and finite dimensional.

1. The group algebra $K[G]$ of a cyclic group G is an integral domain.

2. For every $x \in G$, $x - 1$ is a zero–divisor in $K[G]$.

3. Every non–trivial zero–divisor in $K[G]$ is of the form $x - 1$ or $1 + x + \cdots + x^{m-1}$ for some $x \in G$ whose order is $m \geq 2$.

4. The group algebra of a simple group over a field is simple.

5. One of the isotypical components of $K[G]$ is a minimal left ideal.

6. Any minimal 2–sided ideal of $K[G]$ is generated by a primitive idempotent.

7. The number of isotypical components of $K[G]$ each of which is generated by a primitive idempotent is the order of G^{ab}.

8. If the idempotent 1 in $K[G]$ is primitive, then $G = \{1\}$.

9. $K[G]$ is semi–simple if it has no nilpotent elements.

10. $K[G]$ is semi–simple if $z = \sum_{x \in G} x$ is *not* nilpotent.

11. $K[G]$ is semi–simple if it has no nilpotent ideals.

12. The subset of G–invariants in the regular representation $K[G]$ is K.

13. The subset of G–invariants in the regular representation $K[G]$ is Kz where $z = \sum_{x \in G} x$.

14. The regular representation $K[G]$ is multiplicity free.

15. Any group character of G corresponds to a one dimensional 2–sided ideal of $K[G]$.

16. A representation is determined by its character.

17. An ordinary representation is determined by its character.

18. An irreducible representation is determined by its character.

19. An irreducible representation of an abelian group is a character.

20. Groups with same character tables are mutually isomorphic.

21. The number of distinct characters of G is the order of G^{ab}.

22. A character of norm 1 is irreducible.

23. An ordinary character of norm 1 is irreducible.

24. An ordinary virtual character of norm 1 is an irreducible character.

25. The character of the contragredient dual of an irreducible representation is of norm 1.

26. Irreducibility of a representation is invariant under tensoring with a group character.

27. Norms of the characters of a representation and its dual are equal.

28. Tensor product of irreducible representations is irreducible.

29. Tensor product of ordinary irreducible representations is irreducible.

30. $\text{Hom}_G(V, W)$ is the maximal trivial subrepresentation of $\text{Hom}_K(V, W)$ for any two representations V and W of G.

31. The character of the tensor product of representations is the product of their characters.

32. Any solvable non–abelian (finite) group is of odd order.

33. Any simple solvable group is abelian.

34. If the number of elements of a conjugacy class of G is a positive power of a prime, then G cannot be simple.

35. If the number of elements of a conjugacy class of G is a positive power of a prime, then G is solvable.

36. The character of the dual of a complex character is its complex conjugate.

37. The character of $\text{End}_{\mathbb{C}}(V)$ is real valued for every complex representation V.   ■

Chapter 4

Induced Representations

In this chapter, we shall study an important technique due to Frobenius, called "inducing representations" from subgroups, which enables one to construct representations of a group G in terms of those of its subgroups H. Decomposing induced representations is *not* an easy problem, yet it is one of the most powerful methods for constructing irreducible representations of finite groups.

On the one hand, having constructed a complete set of mutually inequivalent (ordinary) irreducible representations of a group G, Clifford's decomposition theorem coupled with Mackey's irreducible criterion (§§4.4 and 4.6 below), using Frobenius reciprocity (§4.2 below), tells how to get the same for a *normal* subgroup N from those of G. The task becomes relatively simpler if the quotient group G/N is *cyclic*–better still–*of prime order*. In Chapters 6 and 8 below, we need this only for the *best* possible case, namely, for subgroups N of index 2.

On the other hand, if N is an *abelian normal* subgroup of G such that $G = N \bullet H$ is a *semi-direct product* (Ex.(4.8.11) below) of H by N for some subgroup H of G, then the *Wigner–Mackey* method of *little groups* (§4.7 below) shows that all the (ordinary) irreducible representations of G can be constructed from those of N and of certain ("little") subgroups of H. We use this as one of the methods to obtain the irreducibles for the group B_n in Chapter 7 below.

We shall develop the basic facts of the theory of induced representations centering around the Frobenius Reciprocity Theorem, the Decomposition Theorems of Clifford and Mackey, etc. Our exposition is based on [10], [23], [26], [31], [42], [46], etc.

As usual, K stands for a field (*not* necessarily algebraically closed *nor* of characteristic 0 unless specified otherwise) and G a finite group of order $|G|$. All vector spaces considered are over K and finite dimensional.

4.1   Restriction and Induction

4.1.1   Restriction: Let G be a group and $\vartheta : G \to \mathrm{Aut}_K(V)$ be a representation of G on a vector space V over K. Let H be a subgroup of G. The restriction of ϑ to H gives a representation of H on V, called the *restriction of the representation to* H and is denoted by $\mathrm{Res}\,(\vartheta)\!\downarrow_H^G$ or $\vartheta\!\downarrow_H^G$ or ϑ_H or $\mathrm{Res}\,(V)\!\downarrow_H^G$ or $V\!\downarrow_H^G$ or V_H.

The operation of restriction of a $K[G]$–module V to H simply means that V is to be considered as a $K[H]$–module by restriction of scalars.

If χ_ϑ is the character of the representation (V, ϑ) of G, then the character of the restriction (V, ϑ_H) is simply the restriction to H of the map χ_ϑ and is denoted by χ_H.

The following are immediate from definitions.
Additivity of restriction: *If (V_i, ϑ_i) $(i = 1, 2)$ are two representations of G, then* $\mathrm{Res}\,(\vartheta_1 \oplus \vartheta_2)\!\downarrow_H^G = \mathrm{Res}\,(\vartheta_1)\!\downarrow_H^G \oplus \mathrm{Res}\,(\vartheta_2)\!\downarrow_H^G$.
Transitivity of restriction: *If $H \subseteq L$ are subgroups of G, then* $\mathrm{Res}\,(\vartheta)\!\downarrow_H^G = \mathrm{Res}\,\left(\mathrm{Res}\,(\vartheta)\!\downarrow_L^G\right)\!\downarrow_H^L$.

4.1.2   Induction: Given a representation (W, θ) of a subgroup H of G, the base change (or the extension of scalars from $K[H]$) (1.10.13) to $K[G]$ of the $K[H]$–module W, namely, the left $K[G]$–module $K[G] \otimes_{K[H]} W$, is called the *induced module* or the *induced representation* of G, induced by θ, and is denoted by $\mathrm{Ind}\,(\theta)\!\uparrow_H^G$ or $\theta\!\uparrow_H^G$ or θ^G or $\mathrm{Ind}\,(W)\!\uparrow_H^G$ or $W\!\uparrow_H^G$. (Note that the tensor product involved is over a non–commutative ring.)

If χ_θ is the character of the representation (W, θ) of G, then the character of the induced representation is denoted by $\text{Ind}\left(\chi_\theta\right)\uparrow_H^G$ or $\chi_\theta\uparrow_H^G$ or χ_θ^G or $\text{Ind}\left(\chi_W\right)\uparrow_H^G$ or $\chi_W\uparrow_H^G$ or χ_W^G.

4.1.3 Examples: 1. If 1_H is the trivial one–dimensional representation of H, then $\text{Ind}\,(1_H)\uparrow_H^G = K[G] \otimes_{K[H]} K$ is the *permutation representation* of G on the coset space G/H (3.3.5)(3). In particular, **2.** $(K[G], \text{reg}_G) = \text{Ind}\,(\text{reg}_H)\uparrow_H^G = \text{Ind}\,(1_1)\uparrow_1^G$ for subgroups H of G.

The following are equivalent forms of tensor product commuting with direct sums (1.10.5) and transitivity of base change (1.10.14).

Additivity of induction: *If (W_i, θ_i) $(i = 1, 2)$ are two representations of H, then we have* $\text{Ind}\,(\theta_1 \oplus \theta_2)\uparrow_H^G = \text{Ind}\,(\theta_1)\uparrow_H^G \oplus \text{Ind}\,(\theta_2)\uparrow_H^G$.
Transitivity of induction: *If $H \subseteq L$ are subgroups of G, then we have an isomorphism of left $K[G]$–modules*

$$K[G] \underset{K[H]}{\bigotimes} W \simeq K[G] \underset{K[L]}{\bigotimes} \left(K[L] \underset{K[H]}{\bigotimes} W\right),$$

..e., $\text{Ind}\,(\theta)\uparrow_H^G \cong \text{Ind}\left(\text{Ind}\,(\theta)\uparrow_H^L\right)\uparrow_L^G$.

4.1.4 Proposition: *Let (W, θ) be a (finite dimensional) representation of H. Then the dimension of the induced representation $W\uparrow_H^G$ is given by* $\dim_K(W\uparrow_H^G) = [G : H]\dim_K(W)$.

Proof: Start with a left coset decomposition of G relative to H, say $G = \cup_{i=1}^\ell x_i H$ so that $\ell = [G : H]$. We may assume that $x_1 = 1$. Since every element $x \in G$ can be expressed as $x = x_i h$ for unique i and $h \in H$, it follows that $K[G]$ is a *free right $K[H]$–module with basis* $\{x_i\}_{i=1}^\ell$ and hence get a decomposition of right $K[H]$–modules $K[G] = \oplus_{i=1}^\ell x_i K[H]$. Since tensor product commutes with direct sums, we get an isomorphism of K–vector spaces

$$K[G] \underset{K[H]}{\bigotimes} W \simeq \overset{\ell}{\underset{i=1}{\bigoplus}} \left(x_i K[H] \underset{K[H]}{\bigotimes} W\right).$$

We note that $x_i K[H] \otimes_{K[H]} W = x_i \otimes W \simeq W$ (as vector spaces) under the map $x_i y \otimes w \mapsto yw$ for $y \in K[H]$ and $w \in W$. Thus $W\uparrow_H^G \simeq W^\ell$ as vector spaces, implying the result. $\diamond$

4.1.5 Remarks: 1. The summands $x_i \otimes W$ of $W\uparrow_H^G$ are in general *not* left $K[H]$-modules. However, the particular summand

$$x_1 K[H] \bigotimes_{K[H]} W = K[H] \bigotimes_{K[H]} W = 1 \otimes W \simeq W$$

is a left $K[H]$-module. Consequently, the restriction to H of the induced module $W\uparrow_H^G$ contains a copy of W. In symbols, we have a decomposition of $K[H]$-modules that $(W\uparrow_H^G)\downarrow_H^G = W + \cdots$ (which need not be a direct sum of $K[H]$-modules).
2. The summands $x_i \otimes W$ are permuted by G because for $g \in G$, we have $gx_i = x_j h$ for unique j and $h \in H$ and hence $gx_i \otimes W = x_j h \otimes W = x_j \otimes W$. We also note that $g(1 \otimes W) = x_i \otimes W \iff g \in x_i H$.
3. In case H is normal in G, the summands $x_i \otimes W$ are left $K[H]$-modules because for $h \in H$, we have $hx_i = x_i h'$ for some $h' \in H$ and $hx_i \otimes W = x_i \otimes W$.

4.1.6 Theorem: *Let (V, ϑ) be a representation of G and (V, ϑ_H) be its restriction to a subgroup H. Let (W, θ) be a subrepresentation of H in V. Then the following are equivalent.*
1. *(V, ϑ) is induced by (W, θ).*
2. *V is a direct sum of $\{\vartheta(x_i)W\}_{i=1}^{\ell}$ where $X = \{x_1, \cdots, x_\ell\}$ is a complete set of left coset representatives of H in G.*

Proof: If (V, ϑ) is induced by (W, θ), then we have

$$V = \bigoplus_{i=1}^{\ell} x_i \otimes W \text{ and } x_i \otimes W = x_i(1 \otimes W) = \vartheta(x_i)W.$$

Conversely, suppose $V = \oplus_{i=1}^{\ell} \vartheta(x_i)W$. Since W is H-stable, we find that $\vartheta(x_i)W$ depends only on the coset $x_i H$. We know that $K[G]$ is a free right $K[H]$-module with X as a basis. Now look at the map

$$f : K[G] \times W \longrightarrow V, \quad (\sum_{i=1}^{\ell} x_i b_i, w) \mapsto \sum_{i=1}^{\ell} x_i b_i w,$$

for all $b_i \in K[H]$ and $w \in W$ This is balanced (1.10.1) and hence gives rise to a homomorphism $\tilde{f} : K[G] \otimes_{K[H]} W \to V$ which is also left $K[G]$-linear and surjective. For dimension reasons, it has to be an isomorphism, as required.   ◇

4.1.7 Remark: The statement (2) of the theorem above is often taken as the definition of an induced representation for the technical reason that it does not need the concept of "tensor product over non–commutative rings" (1.10.2). However, the existence and uniqueness of an induced representation (which are so obvious from our definition) need to be seen in this version. We leave the details to the reader.

Now we determine the character of an induced representation.

4.1.8 Theorem: *Let* (W, θ) *be a representation of H and* $(W{\uparrow}_H^G, \theta^G)$ *be the induced representation with respective characters* χ_θ *and* χ_θ^G. *Let* $\{x_1, \cdots, x_\ell\}$ *be a complete set of left coset representatives of H in G. For $g \in G$, we have*

$$\chi_\theta^G(g) = \sum_{\substack{1 \le i \le \ell \\ x_i^{-1} g x_i \in H}} \chi_\theta(x_i^{-1} g x_i) = \frac{1}{|H|} \sum_{\substack{x \in G \\ x^{-1} g x \in H}} \chi_\theta(x^{-1} g x).$$

Proof: We have $W{\uparrow}_H^G = \oplus_{i=1}^\ell W_i$ where $W_i = \theta^G(x_i)W$. Choose a basis $\{w_1, \cdots, w_m\}$ for W so that $\{\theta^G(x_i)(w_j) \mid 1 \le i \le \ell, 1 \le j \le m\}$ is a basis of $W{\uparrow}_H^G$. For $g \in G$, let $g x_i = x_{j(i)} h_i$ so that we have $\theta^G(g)W_i = W_{j(i)}$.

By definition, we have $\chi_\theta^G(g) = \text{trace}(\theta^G(g))$. To find this, we have to know the coefficient α_{ik} of $\theta^G(x_i)(w_k)$ in

$$\theta^G(g x_i)(w_k) = \theta^G(x_{j(i)} h_i)(w_k) = \theta^G(x_{j(i)})(\theta(h_i)(w_k))$$

so that $\chi_\theta^G(g) = \sum_{i,k} \alpha_{ik}$. But $\alpha_{ik} = 0$ if $j(i) \ne i$ and when $j(i) = i$ (so that $h_i = x_i^{-1} g x_i$), we see that $\sum_{k=1}^m \alpha_{ik} = \text{trace}(\theta(h_i)) = \chi_\theta(h_i)$. Thus we get that

$$\chi_\theta^G(g) = \sum_{i=1}^\ell \chi_\theta(h_i) = \sum_{\substack{1 \le i \le \ell \\ x_i^{-1} g x_i \in H}} \chi_\theta(x_i^{-1} g x_i),$$

as required. The second equality is an obvious reformulation of the other where each term repeats $|H|$ times. $\Diamond$

4.1.9 Remarks: 1. If we extend the function $\chi_\theta : H \to K$ to G by defining $\tilde{\chi}_\theta : G \to K$ to be

$$\tilde{\chi}_\theta(g) = \begin{cases} \chi_\theta(g) & \text{if } g \in H, \\ 0 & \text{if } g \notin H, \end{cases}$$

then the induced character takes the form

$$\chi_\theta^G(g) = \sum_{1 \le i \le \ell} \tilde{\chi}_\theta(x_i^{-1} g x_i) = \frac{1}{|H|} \sum_{x \in G} \tilde{\chi}_\theta(x^{-1} g x).$$

2. Suppose that the matrix representation of $\theta : H \to \operatorname{Aut}_K(W)$ is given by $\theta(h) = \left(\theta_{ij}(h)\right)$ with respect to a basis $\{w_1, \cdots, w_m\}$ of W. Extend the functions θ_{ij} to $\tilde{\theta}_{ij}$ on G as above and define $\tilde{\theta}(g) = \left(\tilde{\theta}_{ij}(g)\right)$, etc. Then the matrix form of the induced representation θ^G with respect to the basis $\left\{\left((\theta^G(x_i)(w_j))_{1 \le j \le m}\right)_{1 \le i \le \ell}\right\}$ of $W{\uparrow}_H^G$ is given by

$$\theta^G(g) = \left(\tilde{\theta}\left(x_i^{-1} g x_j\right)\right)$$

which is an $\ell \times \ell$ matrix of blocks, each block being an $m \times m$ matrix. The $(ij)^{\text{th}}$ block is then the matrix

$$\begin{aligned}
\tilde{\theta}\left(x_i^{-1} g x_j\right) &= \left(\tilde{\theta}_{pq}\left(x_i^{-1} g x_j\right)\right), \quad 1 \le p, q \le m, \\
&= \begin{cases} \left(\theta_{pq}(x_i^{-1} g x_j)\right) & \text{if } x_i^{-1} g x_j \in H, \\ 0 & \text{if } x_i^{-1} g x_j \notin H. \end{cases}
\end{aligned}$$

Now it is more transparent to find the trace and again get the formula

$$\chi_\theta^G(g) = \sum_{\substack{1 \le i \le \ell \\ x_i^{-1} g x_i \in H}} \chi_\theta(x_i^{-1} g x_i).$$

4.1.10 Corollary: *Suppose* $\dim_K(W) = 1$, *i.e.,* θ *is a character of the subgroup* H. *Then for each* $g \in G$, *the matrix* $\theta^G(g)$ *is a monomial matrix in the sense that in every row and column there is exactly one non–zero entry.*

4.1.11 Monomial representation: A representation (V, ϑ) of a group G is called a *monomial representation* if there is a basis of V with respect to which the matrix of $\vartheta(g)$ is monomial for all $g \in G$.

Representations induced by one–dimensional representations of subgroups are examples of monomial representations (4.1.10). However, a monomial representation of G need not be induced in general. ∎

4.2 Frobenius Reciprocity Theorems

Frobenius reciprocity relates the restriction and induction of representations (resp. characters, resp. class functions) so as to yield information from one to the other (in the case when $K[G]$ is semi-simple, in particular if Char $K = 0$). We assume as usual that K and G are such that $K[G]$ is semi-simple.

Recall (§3.6 above) that on $\mathcal{C}_K(G)$, the space of K-valued class functions on G, there is an inner product $\langle\ ,\ \rangle_G$ with respect to which the characters of the irreducible representations of G form an orthonormal basis (3.6.4).

Given a subgroup H and a K-valued map f on H, let $\tilde{f}$ denote the extension of f to G vanishing outside H (4.1.9)(1). Motivated by the character formula (4.1.9)(2) for an induced representation, we define

$$f^G(g) = \frac{1}{|H|} \sum_{x \in G} \tilde{f}\,(x^{-1}gx), \ \forall\, g \in G,$$

for any map f on H. We note that $f \in \mathcal{C}_K(H) \Rightarrow f^G \in \mathcal{C}_K(G)$.

4.2.1 Theorem (Frobenius reciprocity for class functions): *For $\phi \in \mathcal{C}_K(H)$ and $\Phi \in \mathcal{C}_K(G)$, we have $\langle \phi^G, \Phi \rangle_G = \langle \phi\ ,\ \Phi\ |_H \rangle_H$.*

Proof: Follows just from definitions. ◇

4.2.2 Corollary: *Let (V, ϑ) be a representation of G and (W, θ) be one of H. Then we have $\langle \chi_\theta^G, \chi_\vartheta \rangle_G = \langle \chi_\theta\ ,\ \chi_\vartheta\ |_H \rangle_H$. In particular, if Char $K = 0$ and both V and W are irreducible, then V occurs (upto equivalence) in $\mathrm{Ind}\,(W)\!\uparrow_H^G$ if and only if W occurs in $\mathrm{Res}\,(V)\!\downarrow_H^G$. Furthermore, the multiplicity with which V occurs in $\mathrm{Ind}\,(W)\!\uparrow_H^G$ (3.3.10)(6) is the same as that of W in $\mathrm{Res}\,(V)\!\downarrow_H^G$.*

4.2.3 Theorem (Frobenius reciprocity for representations): *Let (V, ϑ) be a representation of G and (W, θ) be that of a subgroup H. Then we have an isomorphism of K-vector spaces*

$$\mathrm{Hom}_{K[H]}\big(W, \mathrm{Res}\,(V)\!\downarrow_H^G\big) \approx \mathrm{Hom}_{K[G]}\big(\mathrm{Ind}\,(W)\!\uparrow_H^G, V\big).$$

This means that the two operations "Restriction" and "Induction" are adjoint *to each other, as explained below.*

Proof: This is a direct consequence of the "adjointness" of the tensor product and homomorphisms, namely:

Adjointness: *Let R and S be K-algebras. Given a left R-module L, an (S, R)-bimodule M (1.10.8) and a right S-module N, the map*

$$\varphi : \operatorname{Hom}_R\big(L, \operatorname{Hom}_S(M, N)\big) \longrightarrow \operatorname{Hom}_S\big(M \otimes_R L, N\big),$$
$$f \longmapsto \varphi(f)\big(: \; x \otimes y \mapsto f(y)(x), \cdot\, \forall\, x \in M, \; y \in L\big),$$

is an isomorphism of K-modules.

The objects and maps under consideration are all meaningful and the map ψ in the reverse direction

$$g \longmapsto \psi(g)\big(: y \mapsto \psi(g)(y)(: \; x \mapsto g(x \otimes y))\big)$$

is easily seen to be inverse to φ.

Now choosing $R = K[H]$, $S = K[G] = M$, $L = W$ and $N = V$, and using the fact that the map

$$\phi : \operatorname{Hom}_{K[G]}\big(K[G], V\big) \longrightarrow V, \; f \mapsto f(1)$$

is an isomorphism of left $K[G]$-modules, we get an isomorphism of K-modules

$$\operatorname{Hom}_{K[H]}\big(W, \operatorname{Hom}_{K[G]}(K[G], V)\big) \;\; \xrightarrow{\;\simeq\;} \;\; \operatorname{Hom}_{K[G]}\big(K[G] \otimes_{K[H]} W, V\big)$$
$$\| \qquad\qquad\qquad\qquad\qquad\qquad\qquad\qquad \|$$
$$\operatorname{Hom}_{K[H]}\big(W, \operatorname{Res}(V){\downarrow}_H^G\big) \;\; \xrightarrow{\;\simeq\;} \;\; \operatorname{Hom}_{K[G]}\big(\operatorname{Ind}(W){\uparrow}_H^G, V\big)$$

as required.   ◊

The following is another version of the Frobenius reciprocity theorem for representations.

4.2.4 Theorem: *Let (V, ϑ) be a representation of G and (W, θ) that of a subgroup H. Then we have an isomorphism of K-vector spaces*

$$\operatorname{Hom}_{K[G]}\big(V, \operatorname{Ind}(W){\uparrow}_H^G\big) \simeq \operatorname{Hom}_{K[H]}\big(\operatorname{Res}(V){\downarrow}_H^G, W\big).$$

Proof: If $K[G]$ is semi-simple, we have $\operatorname{Hom}_G(M, N) \approx \operatorname{Hom}_G(N, M)$ for all representations M and N of G. Hence the result is the same as (4.2.3). However, in general, the following is a proof from the first principles.

Let $X = \{x_1, \cdots, x_\ell\}$ (with $x_1 = 1$) be a complete set of left coset representatives of H in G. Let $\operatorname{Ind}(W){\uparrow}_H^G = (W{\uparrow}_H^G, \theta^G)$ and write $W{\uparrow}_H^G = \oplus_{i=1}^\ell \theta^G(x_i)W$.

Given a $K[G]$-linear map $f : V \to W{\uparrow}_H^G$; we can write (for each $v \in V$) that $f(v) = \sum_{i=1}^\ell \theta^G(x_i)w_i$ for uniquely determined $w_i \in W$. It is easy to check that we get well-defined K-linear maps $f_i : V \to W$, $f_i(v) = w_i$, $1 \le i \le \ell$.

For $g \in G$, write $gx_i = x_{j(i)}h_i$, for unique $j(i)$ and $h_i \in H$. As i varies between 1 and ℓ so does $j(i)$. Since f is $K[G]$-linear, we have

$$f(\vartheta(g)v) = \theta^G(g)f(v) = \sum_{i=1}^\ell \theta^G\big(x_{j(i)}\big)\big(\theta(h_i)w_i\big).$$

In particular, for $g = x_k^{-1}$, we get that $j(i) = 1 \iff i \neq k$ and so

$$f(\vartheta(x_k^{-1})v) = w_k + \sum_{i \neq k}^\ell \theta^G\big(x_{j(i)}\big)\big(\theta(h_i)w_i\big)$$

which means that $f_1(\vartheta(x_k^{-1})v) = w_k = f_k(v)$ for all k. Hence we can rewrite $f(v)$ as

$(\star)$
$$f(v) = \sum_{i=1}^\ell \theta^G(x_i)f_1\big(\vartheta(x_i^{-1})v\big).$$

Notice that for $h \in H$, we have $hx_i = x_{j(i)}h_i$ with $j(i) = 1 \iff i = 1$ and $h_1 = h$. This gives that

$$f(\vartheta(h)v) = \theta(h)w_1 + \sum_{i=2}^\ell \theta^G\big(x_{j(i)}\big)\big(\theta(h_i)w_i\big).$$

Thus $f_1(\vartheta(h)v) = \theta(h)f_1(v)$, i.e., f_1 is $K[H]$–linear which comes to the same as saying $f_1 \in \mathrm{Hom}_{K[H]}(\mathrm{Res}\,(V){\downarrow}_H^G, W)$.

Define $F : \mathrm{Hom}_{K[G]}\big(V, \mathrm{Ind}\,(W){\uparrow}_H^G\big) \longrightarrow \mathrm{Hom}_{K[H]}\big(\mathrm{Res}\,(V){\downarrow}_H^G, W\big)$ by $F(f) = f_1$. It is clear that F is K–linear and injective because of $(\star)$ above. On the other hand, given $\varphi \in \mathrm{Hom}_{K[H]}\big(\mathrm{Res}\,(V){\downarrow}_H^G, W\big)$, let $\Phi : V \to \mathrm{Ind}\,(W){\uparrow}_H^G$ be defined by the formula $(\star)$, i.e., $\Phi(v) = \sum_{i=1}^{\ell} \theta^G(x_i)\varphi\big(\vartheta(x_i^{-1})v\big)$ so that $F(\Phi) = \varphi$. It remains to check that Φ is $K[G]$–linear. For $g \in G$, we have

$$\begin{aligned}
\Phi(\vartheta(g)v) &= \sum_{i=1}^{\ell} \theta^G(x_{j(i)})\varphi\big(\vartheta(x_{j(i)}^{-1})\vartheta(g)v\big) = \sum_{i=1}^{\ell} \theta^G(x_{j(i)})\varphi\big(\vartheta(h_i x_i^{-1})v\big) \\
&= \sum_{i=1}^{\ell} \theta^G(x_{j(i)}h_i)\varphi\big(\vartheta(x_i^{-1})v\big) = \sum_{i=1}^{\ell} \theta^G(gx_i)\varphi\big(\vartheta(x_i^{-1})v\big) \\
&= \theta^G(g)\Big[\sum_{i=1}^{\ell} \theta^G(x_i)\varphi\big(\vartheta(x_i^{-1})v\big)\Big] = \theta^G(g)\Phi(v). \qquad \blacksquare
\end{aligned}$$

4.3   Conjugate Representations

4.3.1 Conjugate representation: Let H be a subgroup of a group G and (W, θ) be a representation of H. For $g \in G$, let $H^g = gHg^{-1}$ be the g–conjugate of H. The representation (W^g, θ^g) of H^g on the same vector space $W = W^g$, defined by $\theta^g(ghg^{-1}) = \theta(h)$ for all $h \in H$, is called the *conjugate representation* of H, associated to g.

If χ is the character of θ, then the character of θ^g is denoted by χ^g. By definition, we have

$$\chi^g(ghg^{-1}) = \mathrm{trace}(\theta^g(ghg^{-1})) = \mathrm{trace}(\theta(h)) = \chi(h), \ \forall\, h \in H.$$

4.3.2 Remarks: 1. For $x, y \in G$, we have $(W^{xy}, \theta^{xy}) = ((W^y)^x, (\theta^y)^x)$ and $(W, \theta) = ((W^x)^{x^{-1}}, (\theta^x)^{x^{-1}})$. In particular, (W, θ) is H–irreducible if and only if (W^x, θ^x) is H^x–irreducible.

2. It is obvious that $(W^g, \theta^g) = (W, \theta)$ for all $g \in G$ if H is a central subgroup of G. On the other hand, if $g \in G$ normalises H, i.e.,

$H = H^g$, then (W^g, θ^g) is another representation of H, but *not* necessarily equivalent to (W, θ); in fact, even the characters χ and χ^g need not be the same. Conjugate representations occur naturally in many contexts, as in the following.

4.3.3 Proposition: *Suppose (W, θ) is a subrepresentation of the restriction to H of a representation (V, ϑ) of G. Then (W', ϑ') is a representation of H^g where $W' = \vartheta(g)W$ and $\vartheta'(ghg^{-1}) = \vartheta(ghg^{-1})$ for all $h \in H$. Furthermore, the two representations of H^g, namely, (W^g, θ^g) and (W', ϑ') are equivalent.*

Proof: Recall that a subrepresentation of the restriction means that $\vartheta(H)(W) = W$ and $\vartheta\!\downarrow^G_H = \theta$ on W. It is trivial to see that $\vartheta(g)W$ is stable under H^g for all $g \in G$. The map $w \mapsto \vartheta(g)w$ of W^g onto W' is the required isomorphism of H^g–modules since $\theta^g(ghg^{-1}) = \theta(h) = \vartheta(h)$ ($\forall\, h \in H$) and the following diagram is commutative.

$$
\begin{array}{ccc}
W^g & \xrightarrow{\ \vartheta(g)\ } & \vartheta(g)W \\
\Big\downarrow{\scriptstyle\theta^g(ghg^{-1})} & & \Big\downarrow{\scriptstyle\vartheta(ghg^{-1})} \\
W^g & \xrightarrow{\ \vartheta(g)\ } & \vartheta(g)W
\end{array}
\qquad\qquad \diamondsuit
$$

4.3.4 Remark: Suppose $h \in H$. Then the H–modules (W^h, θ^h) and (W, θ) are equivalent (under the natural isomorphism $w \mapsto \theta(h)w$ of W^h onto W).

Note: Under the hypothesis of (4.3.3) or (4.3.4), the speciality of the situation is that the map $w \mapsto \vartheta(g)w$ is meaningful and gives the required isomorphism.

4.3.5 Theorem: *Let W be a representation of a subgroup H of a group G and let $N_G(H)$ be the normaliser of H in G. Then we have the following.*
1. *The set $I_W = I_\theta = \{g \in N_G(H) \mid (W^g, \theta^g) \cong (W, \theta)\}$ is a subgroup containing H, called the inertia group of (W, θ).*
2. *For $g, h \in N_G(H)$, we have $(W^g, \theta^g) \cong (W^h, \theta^h) \iff g^{-1}h \in I_\theta$.*

Proof: 1. Let $g \in I_\theta$, say $\lambda : W \to W^g$ is an isomorphism of H–

modules so that (for all $x = gyg^{-1} \in gHg^{-1}$), the diagram

$$
\begin{array}{ccc}
W & \xrightarrow{\ \lambda\ } & W^g \\
\downarrow{\scriptstyle\theta(x)} & & \downarrow{\scriptstyle\theta^g(x)=\theta(y)} \\
W & \xrightarrow{\ \lambda\ } & W^g
\end{array}
$$

is commutative, i.e., $\lambda \circ \theta(x) = \theta(y) \circ \lambda$. This gives the relation that $\lambda \circ \theta^{g^{-1}}(y) = \theta(y) \circ \lambda$ for all $y = g^{-1}xg \in g^{-1}Hg$, i.e., $(W^{g^{-1}}, \theta^{g^{-1}}) \cong (W, \theta)$ and so $g^{-1} \in I_\theta$.

Now suppose that $g, h \in I_\theta$. Then we have $(W, \theta) \cong (W^g, \theta^g) \cong ((W^h)^g, (\theta^h)^g) = (W^{gh}, \theta^{gh})$ which implies that $gh \in I_\theta$ and hence I_θ is a subgroup.

The second part is immediate since $(W^g, \theta^g) \cong (W^h, \theta^h) \Rightarrow (W, \theta) \cong ((W^h)^{g^{-1}}, (\theta^h)^{g^{-1}})$ and hence $g^{-1}h \in I_\theta$, as required. $\diamond$

4.3.6 Proposition: *Let (W, θ) be a representation of H. Then the representations induced by all the conjugates of W are equivalent, i.e., we have* $\mathrm{Ind}\,(W){\uparrow}_H^G \cong \mathrm{Ind}\,(W^g){\uparrow}_{H^g}^G$ *for all $g \in G$ (as $G - modules$).*

Proof: It is easy to see that the natural map
$$\varphi_g : K[G] \otimes_{K[H]} W \xrightarrow{\ \cong\ } K[G] \otimes_{K[H^g]} W^g, \quad \alpha \otimes w \mapsto \alpha g^{-1} \otimes w$$
is an isomorphism of left $K[G]$–modules. $\diamond$

4.3.7 Corollary: *Let $\mathrm{Char}\, K = 0$. Let (V, ϑ) be an irreducible representation of G and (W, θ) an irreducible representation of a normal subgroup H. If W occurs (upto equivalence) in $\mathrm{Res}\,(V){\downarrow}_H^G$, then all the conjugates of W also occur, as clear from (4.3.3).*

4.3.8   Remark: Clifford's decomposition theorem ((4.4.1) below) says that (4.3.7) above is true (i.e., all the conjugates occur with the same multiplicity) (3.3.10)(6) _without_ the _assumption_ that $\mathrm{Char}\, K = 0$.   ■

4.4   Clifford's Decomposition Theorem

Let N be a *normal* subgroup of a group G. For a representation (W, θ) of N, let I_W be the *inertia* group of W (4.3.5). Let $\{z_1, \cdots, z_m\}$ (with

$z_1 = 1$) be a complete set of left coset representatives of I_W in G. By definition of I_W, note that $\{(W^{z_i}, \theta^{z_i})\}_{i=1}^{m}$ is a complete set of mutually inequivalent conjugate representations of N. Each element $g \in G$ permutes these conjugates under the isomorphism

$$(W^{z_i})^g \cong W^{gz_i} \cong W^{z_{j(i)}} \quad \text{where} \quad gz_i = z_{j(i)}y_i, \ y_i \in I_W.$$

We keep this notation in what follows.

4.4.1 Theorem (Clifford): *Let K be a field of arbitrary character-istic. Let (V, ϑ) be an irreducible representation of G. Suppose (W, θ) is an irreducible subrepresentation of N in $\mathrm{Res}\,(V){\downarrow}_N^G$. Then we have the following.*
1. *$\mathrm{Res}\,(V){\downarrow}_N^G$ is a direct sum of conjugates of W (hence is a semi-simple $K[N]$-module). Upto equivalence, each conjugate occurs with the same multiplicity, say e (3.3.10)(6).*
2. *Let V_i be the isotypical component of type W^{z_i} (2.2.4). Then*

$$V_i = \underbrace{W^{z_i} \oplus \cdots \oplus W^{z_i}}_{e \text{ copies}} \quad \text{and} \quad \mathrm{Res}\,(V){\downarrow}_N^G = \bigoplus_{i=1}^{m} V_i = \left(\bigoplus_{i=1}^{m} W^{z_i}\right)^e.$$

In particular, the number of isotypical components of $\mathrm{Res}\,(V){\downarrow}_N^G$ is the index of the inertia group of W.
3. *G transitively permutes the isotypical components of $\mathrm{Res}\,(V){\downarrow}_N^G$.*
4. *The isotypical component V_1 (of type W) is a $K[I_W]$-module and $V \cong \mathrm{Ind}\,(V_1){\uparrow}_{I_W}^G$ as $K[G]$-modules.*

Proof: Since N is normal and W is irreducible, W^g is N-simple for every $g \in G$. Since $\sum_{g \in G} \vartheta(g)W$ is a non-zero G-submodule of V, we get that $V = \sum_{g \in G} \vartheta(g)W$ by G-simplicity of V. Since $\vartheta(g)W \cong W^g$ as N-modules, we get that V is a sum of simple N-modules and hence a direct sum of a suitable (possibly smaller) collection of conjugates of W. Notice that V contains all conjugates of W as well (since it is a sum of all of them). Since $\{W^{z_i}\}_{i=1}^{m}$ is a complete set of mutually inequivalent conjugate representations of N, it follows that $\mathrm{Res}\,(V){\downarrow}_N^G = \oplus_{i=1}^{m}V_i$, the direct sum of its isotypical components where V_i is isotypical of type W^{z_i}. That G permutes these components is obvious since it does so on the W^{z_i}'s, as noted above.

Since $\vartheta(z_i)V = V$, it follows that the multiplicity of W is the same as that of $\vartheta(z_i)W \cong W^{z_i}$. Finally, since $V = \oplus_{i=1}^m \vartheta(z_i)V_1$ and V_1 is an I_W–module, it follows (by (4.1.6) above) that $V \cong \operatorname{Ind}(V_1){\uparrow}_{I_W}^G$.　◇

4.4.2 Corollary: *The restriction of an irreducible representation (V, ϑ) of G to N is isotypical if and only if the inertia group of any N-simple submodule of V is G.*

4.4.3 Corollary: *Let (V, ϑ) be an irreducible representation of G. Then we have the following.*
1. *Either $\vartheta{\downarrow}_N^G$ is isotypical, (in which case, if N is also abelian, then $\vartheta(a)$ is a scalar multiplication for all $a \in N$),*
2. *or else, there exists a subgroup H such that $N \subseteq H \neq G$ and an irreducible representation (W, θ) of H with $\vartheta = \theta{\uparrow}_H^G$.*　∎

4.5　Mackey's Irreducibility Criterion

To begin with, let us recall some elementary facts about double cosets. Let P and Q be subgroups of a group G. On the set of left cosets $G/Q = \{gQ \mid g \in G\}$ of Q in G, P acts by left multiplication.

4.5.1 Double cosets: The P–orbits in G/Q are called (P, Q)–*double cosets* in G and they are of the form PxQ, for $x \in G$. (Equivalently, the Q–orbits in $P\backslash G = \{Pg \mid g \in G\}$ for right multiplication are the same (P, Q)–double cosets.)

For $x, y \in G$, it is trivial to see that $PxQ = PyQ$ if and only if $px = yq$ for some $p \in P$ and $q \in Q$. The set of distinct (P, Q)–double cosets, denoted by $P\backslash G/Q$, gives a partition of G. By a complete set of (P, Q)–double coset representatives, we mean a subset X of G such that the double cosets $\{PxQ \mid x \in X\}$ are distinct and $G = \bigcup_{x \in X} PxQ$.

When there is no danger of confusion, we denote a complete set of (P, Q)–double coset representatives also by $P\backslash G/Q$.

In case $P = Q$, a (P, P)–double coset is simply called a *double coset of P.* In case P is normal, double cosets of P are nothing but the usual cosets.

4.5.2 Proposition: *For $x \in G$, let Y_x be a complete set of left coset representatives of $P \cap Q^x$ in P where $Q^x = xQx^{-1}$. Then $PxQ = \cup_{y \in Y_x} yxQ$ (a disjoint union) and in particular, we have $|PxQ| = [P : (P \cap Q^x)]|Q|$.*

Proof: Obvious from the fact that for $y, z \in P$, we have $yxQ = zxQ$ if and only if $y^{-1}z \in P \cap Q^x$. $\diamond$

Note: The subset Y_x, as above, depends only on the double coset PxQ but not on its representative x (contrary to what the notation suggests). If X is a complete set of (P, Q)–double.coset representatives, then it follows that $Y = \{yx \mid y \in Y_x, x \in X\}$ is a complete set of left coset representatives of Q in G.

4.5.3 Subgroup Theorem: *Let P and Q be subgroups of a group G. Let (W, θ) be a representation of Q. Let $P\backslash G/Q$ be a complete set of (P, Q)–double coset representatives. Then we have a decomposition of P–modules*

$$\operatorname{Res}\left(\operatorname{Ind}(W){\uparrow_Q^G}\right){\downarrow_P^G} \cong \bigoplus_{x \in P\backslash G/Q} \operatorname{Ind}\left(\operatorname{Res}(W^x){\downarrow_{P \cap Q^x}^{Q^x}}\right){\uparrow_{P \cap Q^x}^P}.$$

Proof: 1. For $x \in X = P\backslash G/Q$, let Y_x and Y be as above. By (4.1.6) above, we can write $W \uparrow_Q^G = \oplus_{z \in Y} \theta^G(z)W$. Let $W \uparrow_Q^G (x) = \oplus_{y \in Y_x} \theta^G(yx)W$ (which is obviously P–stable). Thus we have a decomposition of $\operatorname{Res}\left(W{\uparrow_Q^G}\right){\downarrow_P^G} = \oplus_{x \in X} W{\uparrow_Q^G}(x)$, as P–modules. Now look at each of the summands in this decomposition.

2. By (4.3.3) above, we have $\theta^G(x)W \cong W^x$ as Q^x–modules and in particular as $(P \cap Q^x)$–modules (on restriction).

3. The stabiliser of $\theta^G(x)W$ in P, namely, $\{p \in P \mid \theta^G(px)W = \theta^G(x)W\}$ is a subgroup of P and obviously contains $P \cap Q^x$. A simple verification, using the fact that the stabiliser of W in $W{\uparrow_Q^G}$ is precisely Q, shows that it is in fact equal to $P \cap Q^x$. But then by definition of $W{\uparrow_Q^G}(x)$, we find that

$$W{\uparrow_Q^G}(x) = \bigoplus_{y \in Y_x} \theta^G(yx)W = \bigoplus_{y \in Y_x} \theta^G(y)\theta^G(x)W$$

with Y_x being a complete set of left coset representatives of $P \cap Q^x$ in P. Now it follows by (4.1.6) and (2) above, that (as P–modules)

$$W\uparrow_Q^G(x) \cong \mathrm{Ind}\left(\theta^G(x)W\right)\uparrow_{P\cap Q^x}^P = \mathrm{Ind}\left(\mathrm{Res}\,(W^x)\downarrow_{P\cap Q^x}^{Q^x}\right)\uparrow_{P\cap Q^x}^P. \quad \Diamond$$

4.5.4 Corollary: *Let H be a subgroup of G and (W,θ) a representation of H. Then we have a decomposition of H–modules*

$$\mathrm{Res}\left(\mathrm{Ind}\,(W)\uparrow_H^G\right)\downarrow_H^G \;\cong\; \bigoplus_{x\in H\backslash G/H} \mathrm{Ind}\left(\mathrm{Res}\,(W^x)\downarrow_{H\cap H^x}^{H^x}\right)\uparrow_{H\cap H^x}^H$$

$$= \;\; W \bigoplus_{\substack{x\in H\backslash G/H \\ x\notin H}} \mathrm{Ind}\left(\mathrm{Res}\,(W^x)\downarrow_{H\cap H^x}^{H^x}\right)\uparrow_{H\cap H^x}^H.$$

4.5.5 Corollary: *Let H be a normal subgroup of G and (W,θ) a representation of H. Then we have a decomposition of H–modules*

$$\mathrm{Res}\left(\mathrm{Ind}\,(W)\uparrow_H^G\right)\downarrow_H^G \;\cong\; \bigoplus_{x\in G/H} W^x.$$

4.5.6 Corollary: *Let H be a subgroup of G and (W,θ) a representation of H. Then we have a decomposition of H–modules*

$$\mathrm{Res}\left(\mathrm{Ind}\,(1_H)\uparrow_H^G\right)\downarrow_H^G \;\cong\; \bigoplus_{x\in H\backslash G/H} \mathrm{Ind}\,(1_{H\cap H^x})\uparrow_{H\cap H^x}^H.$$

In particular, if H is normal, then $\mathrm{Res}\left(\mathrm{Ind}\,(1_H)\uparrow_H^G\right)\downarrow_H^G$ is a trivial representation of dimension $[G:H]$.

In the rest of this chapter, we assume that Char $K = 0$.

4.5.7 Theorem (Mackey): *Let H be a subgroup of G and (W,θ) a representation of H. Then the following are equivalent.*
1. *The induced representation $\mathrm{Ind}\,(W)\uparrow_H^G$ is irreducible.*

2. $\begin{cases} \text{(a)} & W \text{ is irreducible and} \\ \text{(b)} & \mathrm{Res}\,(W)\downarrow_{H\cap H^x}^H \text{ and } \mathrm{Res}\,(W^x)\downarrow_{H\cap H^x}^{H^x} \text{ have no common} \\ & \text{irreducible components for any } x\notin H. \end{cases}$

Proof: Recall that an ordinary representation is irreducible if and only if its character is of norm 1.

(1) $\Rightarrow$ (2): We have

$$
\begin{aligned}
1 &= \langle \chi_\theta^G, \chi_\theta^G \rangle_G = \langle \chi_\theta, \mathrm{Res}\left(\chi_\theta^G\right) \downarrow_H^G \rangle_H \quad \text{(by reciprocity)} \\
&= \sum_{x \in H \backslash G / H} \langle \chi_\theta, \mathrm{Ind}\left(\mathrm{Res}\left(\chi_\theta^x\right) \downarrow_{H \cap H^x}^{H^x}\right) \uparrow_{H \cap H^x}^H \rangle_H \\
&= \sum_{x \in H \backslash G / H} \langle \mathrm{Res}\left(\chi_\theta\right) \downarrow_{H \cap H^x}^H, \mathrm{Res}\left(\chi_\theta^x\right) \downarrow_{H \cap H^x}^{H^x} \rangle_{H \cap H^x} \\
&= d_1 + \sum_{x \notin H,\ x \in H \backslash G / H} d_x,
\end{aligned}
$$

where $d_x = \langle \mathrm{Res}\left(\chi_\theta\right) \downarrow_{H \cap H^x}^H, \mathrm{Res}\left(\chi_\theta^x\right) \downarrow_{H \cap H^x}^{H^x} \rangle_{H \cap H^x}$ hence $d_1 = 1$ and $d_x = 0$ for all $x \notin H$ (since $d_1 \geq 1$ and $d_x \geq 0$, $\forall\ x$, etc.)

(2) $\Rightarrow$ (1): Retrace the equations above. $\qquad \Diamond$

4.5.8 Corollary: *Let H be normal. Then $\mathrm{Ind}\,(W) \uparrow_H^G$ is irreducible if and only if W is irreducible and the inertia group of W is H (i.e., all the conjugates of W are mutually inequivalent).*

4.5.9 Corollary: *Let (K, θ) be a character of the group H. Then $\mathrm{Ind}\,(\theta) \uparrow_H^G$ is irreducible $\iff \mathrm{Res}\,(\theta) \downarrow_{H \cap H^x}^H \neq \mathrm{Res}\,(\theta^x) \downarrow_{H \cap H^x}^{H^x}$ for all $x \notin H$. In particular, if H is also normal, then $\mathrm{Ind}\,(\theta) \uparrow_H^G$ is irreducible $\iff \theta \not\cong \theta^x$ for all $x \notin H$.* ∎

4.6 Subgroups of Index 2

In this section, we shall apply the above techniques to the very special case when the subgroup H *is of index* 2 in G (which is all that is needed in Chapters 6 and 8). More generally, we assume that N is a normal subgroup of G of index q. For $x \in G$, let C_x be the conjugacy class through x. It is obvious that $C_x \subseteq N$ or $C_x \cap N = \emptyset$ according as $x \in N$ or $x \notin N$. It follows that for $x \in N$, C_x is a union of conjugacy classes of N.

If $q = 2$ and $x \in N$, it is quite simple to distinguish whether or not C_x splits. Let $C_N(x)$ and $C_G(x)$ be the centralisers of x in N and G respectively. We have $C_N(x) = N \cap C_G(x)$ which is a subgroup of index *at most* 2 in $C_G(x)$.

4.6.1 Proposition: *With notation as above, we have the following.*
1. C_x *remains a conjugacy class of N if and only if $C_N(x) \neq C_G(x)$,*
i.e., x commutes with an element outside N.
2. *When $C_N(x) = C_G(x)$, then C_x splits into exactly 2 conjugacy*
classes of N, both of equal cardinality.

Proof: 1. Suppose C_x remains a conjugacy class of N. We have $|C_x|$
$= [G : C_G(x)] = [N : C_N(x)]$ which is not possible if $C_N(x) = C_G(x)$.
Conversely, let x commute with a $t \notin N$. Take $y, z \in C_x$ so that there
are $\sigma, \tau \in G$ such that $y = \sigma x \sigma^{-1}$ and $x = \tau z \tau^{-1}$. But

$$y = (\sigma\tau)z(\tau^{-1}\sigma^{-1}) = (\sigma t\tau)z(\sigma t\tau)^{-1} \text{ with } (\sigma\tau)^{-1}(\sigma t\tau) = \tau^{-1}t\tau \notin N$$

which means that either $\sigma\tau$ or $\sigma t\tau$ is in N.

2. This is trivial because we have

$$[N : C_N(x)] = \frac{1}{2}[G : N][N : C_N(x)] = \frac{1}{2}[G : C_G(x)] = \frac{1}{2}|C_x|. \quad \diamond$$

4.6.2· We go back to the case of arbitrary q. Let (V, ϑ) be an irreducible representation of G and (W, θ) an N–irreducible component
of Res$(V){\downarrow}_N^G$ of multiplicity e (3.3.10)(6). Let I_W be the inertia group
of W. Let $\{z_1, \cdots, z_{q'}\}$ (with $z_1 = 1$) be a complete set of coset representatives of I_W in G so that we have $q' = [G : I_W]$. By Clifford's
Theorem (4.4.1) above, we have Res$(V){\downarrow}_N^G \cong \left(\oplus_{i=1}^{q'} W^{z_i}\right)^e$ so that we
have $\dim_K(V) = eq' \dim_K(W)$. Furthermore, by reciprocity (since
Char $K = 0$), the multiplicity of V in Ind$(W){\uparrow}_N^G$ is exactly e and so
we get that $q \dim_K(W) \geq e \dim_K(V)$, leading to the inequalities:

$$q \dim_K(W) \geq e \dim_K(V) = e^2 q' \dim_K(W) \Rightarrow q \geq e^2 q'.$$

Now we record the following special cases (some of which are
overlapping).

4.6.3 *If all the conjugates of W are inequivalent, then $q = q'$ and*
hence $e = 1$, i.e., Res$(V){\downarrow}_N^G$ *is multiplicity free* (3.3.10)(8).

4.6.4 *Suppose that q is a prime.* Then the inertia group of W is
either N or G, i.e., the conjugates of W are pairwise inequivalent or

all are mutually equivalent (according as $q' = q$ or 1). Hence we have

$$\begin{cases} e = 1 & \text{if the conjugates of } W \text{ are pairwise inequivalent or} \\ q \geq e^2 & \text{if all the conjugates of } W \text{ are mutually equivalent.} \end{cases}$$

In particular, we have the following in the case when $q = 2$.

4.6.5 Theorem: *Let N be of index 2 in G. Then*
1. <u>either</u> $\text{Res}\,(V){\downarrow}_N^G = W^0$ *is itself irreducible and hence self-conjugate* (*in this case,* $\text{Ind}\,(W^0){\uparrow}_N^G = V \oplus V'$ *where V' is irreducible and inequivalent to V, but* $\text{Res}\,(V){\downarrow}_N^G \cong \text{Res}\,(V'){\downarrow}_N^G$),
2. <u>or else</u>, $\text{Res}\,(V){\downarrow}_N^G$ *is reducible with* $\text{Res}\,(V){\downarrow}_N^G = W^+ \oplus W^-$ *being a sum of two inequivalent irreducible components each of half the dimension of V, where $W^+ = W$ and $W^- = \vartheta(x)W$ for any $x \in G, x \notin N$. Moreover, by Mackey's irreducibility criterion (4.5.8),* $\text{Ind}\,(W^+){\uparrow}_N^G \cong V \cong \text{Ind}\,(W^-){\uparrow}_N^G$.
3. *Every irreducible representation of N is equivalent to one of the three kinds, namely, W^0 or W^+ or W^-. In other words, as V varies over the irreducibles of G, a complete set of inequivalent irreducible representations of N is given by*
$$\{W^0 \mid W^0 = \text{Res}\,(V){\downarrow}_N^G\} \cup \{W^\pm \mid \text{Res}\,(V){\downarrow}_N^G = W^+ \oplus W^-\}.$$
4. *For every irreducible representation W of N, $\text{Ind}\,(W){\uparrow}_N^G$ is multiplicity free (3.3.10)(8) and has at the most two irreducible components. Furthermore, every irreducible representation of G occurs as a component of $\text{Ind}\,(W){\uparrow}_N^G$ for some irreducible representation W of N.* ∎

4.7 Wigner–Mackey Method of Little Groups

In this section, we assume that $G = N \bullet H$ is a *semi-direct product* of a group H by an *abelian* group N (Ex.(4.8.11) below). We shall determine the irreducible representations of G, in terms of those of N (i.e., group characters of N) and those of *suitable* subgroups, the so called *little subgroups*, of H. (As stated already, the base field K is assumed to be algebraically closed and of characteristic 0.)

4.7.1 Let the action of H on N be given by a homomorphism of groups $\phi : H \to \text{Aut}_{\mathbb{Z}}(N)$, $x \mapsto \phi_x$, $\forall\, x \in H$. We shall sometimes write $\phi_x(a) = x \cdot a$, for $x \in H$ and $a \in N$.

Recall that every element $z \in G$ can be written uniquely as $z = ax$ for $a \in N$ and $x \in H$ and that N is a normal subgroup of G, etc. We have an exact sequence of groups

$$\{1\} \longrightarrow N \longrightarrow G \longrightarrow G/N \approx H \longrightarrow \{1\}.$$

Since N is normal, G acts on N by inner conjugation. Since N is also abelian, this action of G goes down to an action of $G/N = H$. But this induced action of H coincides with the action ϕ, we started with.

4.7.2 Since N is abelian, the set $\text{Irr}_K(N)$ of all inequivalent irreducible representations of N is nothing but the group $\widetilde{N}$ of all group characters of N. The action of G on N induces one on $\widetilde{N}$, namely,

$$(x \cdot \chi)(a) = \chi(x^{-1}ax), \ \forall\, x \in G,\ \chi \in \widetilde{N} \text{ and } a \in N.$$

We note that the character $x \cdot \chi$ of N is the same as χ^x, the *conjugate* to χ conjugated by $x \in G$. Furthermore, we see that the *inertia* group I_χ of χ is the same as the *isotropy* subgroup of G at χ and $N \subseteq I_\chi$. In fact, if $H_\chi = I_\chi \cap H$ is the isotropy subgroup of H at χ, then $I_\chi = N \bullet H_\chi$ is the semi–direct product of H_χ by N.

4.7.3 Extend the character χ of N to one of I_χ, denoted again by χ, by defining $\chi(ax) = \chi(a)$ for all $a \in N$ and $x \in H_\chi$, which is well–defined since H_χ is the isotropy at χ.

4.7.4 Since $I_\chi/N \approx H_\chi$, any representation (W, ϱ) of H_χ can be naturally treated as a representation of I_χ. Furthermore, the irreducibility or otherwise of (W, ϱ) for H_χ or for I_χ is the same.

4.7.5 Let $\chi \in \widetilde{N}$ and hence $\chi \in \widetilde{I_\chi}$. Let (W_ϱ, ϱ) be an irreducible representation of H_χ (and hence of I_χ). Then we have the tensor product representation $\chi \otimes \varrho$ of I_χ on W which is irreducible (by Ex.(3.11.16) above).

4.7.6 Decompose $\widetilde{N}$ into G–orbits (which are the same as the H–

orbits) and choose a complete set $\{\chi_j \in \widetilde{N} \mid 1 \le j \le r\}$ of representatives of these orbits.

Write $I_j = I_{\chi_j}$ and $H_j = H_{\chi_j}$ so that $I_j = N \bullet H_j$, $1 \le j \le r$, etc. The "*little subgroups*" of H that we are interested in are the subgroups H_j, $1 \le j \le r$. Let $\theta_{j\varrho} = (\chi_j \otimes \varrho)\uparrow_{I_j}^{G}$ (which is a representation of G of dimension $[G : I_j] \dim_K(W_\varrho)$). The main theorem of this section is the following.

4.7.7 Theorem (Wigner–Mackey): *A complete set of mutually inequivalent irreducible representations of G is given by*

$$\mathrm{Irr}_K(G) = \{\theta_{j\varrho} = (\chi_j \otimes \varrho)\uparrow_{I_j}^{G} \mid \varrho \in \mathrm{Irr}_K(H_j) \text{ and } 1 \le j \le r\},$$

where $\mathrm{Irr}_K(H_j)$ is a complete set of mutually inequivalent irreducible representations of H_j. In other words, we have
1. $\theta_{j\varrho}$ *is irreducible for all irreducible representations ϱ of H_j,*
2. $\theta_{j\varrho} \cong \theta_{k\vartheta} \iff j = k$ and $\varrho \cong \vartheta$ *and*
3. *every irreducible representation of G is equivalent to $\theta_{j\varrho}$ for some j and $\varrho \in \mathrm{Irr}_K(H_j)$.*

Proof: We prove **(1)** by using Mackey's irreducibility criterion (4.5.7) above. To do this, fix a j and take $x \notin I_j = N \bullet H_j$. Let us write

$$I = I_j \text{ and } I_x = I \cap (I)^x = I \cap (xIx^{-1}).$$

Note that $N \subseteq I_x$. We have to show that $(\chi_j \otimes \varrho)\downarrow_{I_x}^{I}$ and $(\chi_j \otimes \varrho)^x \downarrow_{I_x}^{I^x}$ do not have a common irreducible component. If they do have a common component, on further restriction to N, we find that $(\chi_j \otimes \varrho)\downarrow_{N}^{I}$ and $(\chi_j \otimes \varrho)^x \downarrow_{N}^{I^x}$ should have a common component. But this is not possible because we have the following.
(i) $(\chi_j \otimes \varrho)\downarrow_{N}^{I} = \dim_K(W_\varrho) \chi_j$, a multiple of χ_j since $\varrho\downarrow_{N}^{I}$ is trivial as ϱ is extended from H_j to I by means of $I \to I/N \approx H_j$,
(ii) $(\chi_j \otimes \varrho)^x \downarrow_{N}^{I^x}$ is a multiple of χ_j^x since $(\chi_j \otimes \varrho)^x = \chi_j^x \otimes \varrho^x$ and $I^x/N \approx H_j^x$, etc., and (iii) $\chi_j \ne \chi_j^x$ for $x \notin I_j$.

We prove **(2)** in two parts. Let V be the representation space of $\theta_{j\varrho}$.
(i) By (4.5.4) above, we observe that $(\theta_{j\varrho})\downarrow_{N}^{G}$ is a sum of the characters of N belonging only to the H–orbit $H \cdot \chi_j$ ($= G \cdot \chi_j$) and hence j is

determined by θ_{j_ϱ}.

(ii) Now look at the N–eigensubspace V_j of V with weight χ_j, i.e., $V_j = \{v \in V \mid \theta_{j_\varrho}(a)v = \chi_j(a)v; \ \forall \, a \in N\}$. It is trivial to check that V_j is H_j–stable. In fact, since $G/I_j = H/H_j$ and

$$V = \bigoplus_{x \in G/I_j} \theta_{j_\varrho}(x)(\chi_j \otimes W_\varrho) = \bigoplus_{x \in H/H_j} \theta_{j_\varrho}(x)(V_j),$$

we find that $V_j \cong W_\varrho$ as H_j–modules and so θ_{j_ϱ} determines ϱ.

To prove **(3)**, start with an irreducible representation (V, θ) of G. Decompose the N–module $V \downarrow_N^G$ into isotypical components, say $V = \bigoplus_{\chi \in \widetilde{N}} V_\chi$, where $V_\chi = \{v \in V \mid \theta(a)v = \chi(a)v, \ \forall \, a \in N\}$. Since $V \neq (0)$, $V_{\chi_j} \neq (0)$ for at least one j. Fix such a j. Note that $\theta(x)(V_\chi) = V_{\chi^x}$ for all $\chi \in \widetilde{N}$ and $x \in G$. In particular, we get that V_{χ_j} is H_j–stable. Let W be an irreducible H_j–submodule of V_{χ_j}. Note that W, being a subspace of V_{χ_j}, is also N–stable. Moreover, these H_j and N actions on W commute. Hence W is an irreducible I_j–module which can be factored as $\chi_j \otimes \varrho$ where ϱ is the representation of I_j on W extended from the quotient $I_j/N = H_j$, etc. Thus we get that $\chi_j \otimes \varrho$ occurs in $\theta \downarrow_{I_j}^G$ and hence by reciprocity (4.2.3) above, we obtain that θ_{j_ϱ} occurs in θ. But then equality results by irreducibility of θ, as required. $\diamond$

4.7.8 Corollary: *Let ξ be a group character of G such that the following are true.*
1. *ξ is trivial on H,*
2. *$(\xi \downarrow_N^G)\chi_j$ is in the G-orbit of χ_k and*
3. *the isotropy subgroups H_j and H_k of H (at χ_j and χ_k) are conjugates in H by some $h \in H$.*
Then we have $\theta_{j_\varrho} \otimes \xi \cong \theta_{k_{\varrho^h}}, \ \forall \, \varrho \in \mathrm{Irr}_K(H_j)$.

4.7.9 Corollary (Mackey): *A character χ of N is the restriction of one of G $\iff$ all the conjugates χ^g of χ are equivalent.* ∎

4.8 Exercises

In what follows, G stands for a finite group and H a subgroup of G. All representations considered are supposed to be over a fixed ground field K (which is algebraically closed and of characteristic 0) and finite dimensional.

1. Let (V, θ) be a representation of H. Let $\overline{V} = \{f : G \to V \mid f(hg) = \theta(h)f(g), \forall h \in H \text{ and } g \in G\}$. Under pointwise addition and usual scalar multiplication of maps, show that $\overline{V}$ is a vector space over K. Let $\overline{\theta}$ be the representation of G on $\overline{V}$ defined by $(\overline{\theta}(g)f)(x) = f(xg)$, $\forall x, g \in G$ and $f \in \overline{V}$. Show that $(\overline{V}, \overline{\theta}) \cong V\uparrow_H^G$.

2. Show that induction commutes with base change, i.e., if $L \supseteq K$ is an extension of fields, then (as $L[G]$–modules) show that $(V\uparrow_H^G) \otimes_K L \cong (V \otimes_K L)\uparrow_H^G$ for all $K[H]$–modules V.

3. Let (V, θ) be a representation of H and V_i ($i = 1, 2$) be H–submodules of V. As G–modules, show that
   (i) $V_1 \subseteq V_2$ if and only if $V_1\uparrow_H^G \subseteq V_2\uparrow_H^G$,
   (ii) $(V_1 + V_2)\uparrow_H^G \cong (V_1\uparrow_H^G) + (V_2\uparrow_H^G)$ and
   (iii) $(V_1 \cap V_2)\uparrow_H^G \cong (V_1\uparrow_H^G) \cap (V_2\uparrow_H^G)$.

4. Let $U \xrightarrow{\alpha} V \xrightarrow{\beta} W$ be a sequence of H–modules. Then show that
   (i)   the sequence $(\star) : 0 \longrightarrow U \xrightarrow{\alpha} V \xrightarrow{\beta} W \longrightarrow 0$ of H–modules is *exact* if and only if the induced sequence of G–modules
   $(\star)\uparrow_H^G : 0 \longrightarrow U\uparrow_H^G \xrightarrow{\alpha \otimes 1} V\uparrow_H^G \xrightarrow{\beta \otimes 1} W\uparrow_H^G \longrightarrow 0$ is exact,
   (ii) $(\star)$ is *split exact* if and only if $(\star)\uparrow_H^G$ is so and
   (iii) $(V/U)\uparrow_H^G \cong (V\uparrow_H^G)/(U\uparrow_H^G)$.

5. Show that induction commutes with the contragredient dual (3.10.10), i.e., $(V^\star)\uparrow_H^G \cong \left(V\uparrow_H^G\right)^\star$.

6. Let V be a G–module and W be an H–module. Then show that $(\mathrm{Hom}_K(W, V))\uparrow_H^G \cong \mathrm{Hom}_K(W\uparrow_H^G, V)$ and $\left(\mathrm{Hom}_K(V\downarrow_H^G, W)\right)\uparrow_H^G \cong \mathrm{Hom}_K(V, W\uparrow_H^G)$.

7. Let (V, θ) be a representation of H. Show that

$$\mathrm{Ker}(\theta\uparrow_H^G) = \bigcap_{g \in G} g\, \mathrm{Ker}(\theta)g^{-1}.$$

8. For a group character χ of H, let $\chi(H) = \sum_{h \in H} \chi(h^{-1})h \in K[G]$. Show that the natural map $\chi\uparrow_H^G \to K[G]\chi(H)$, $1 \otimes 1 \mapsto \chi(H)$, is an isomorphism of G-modules. (Use Ex.(3.11.8) above.)

9. **Theorem (Osima):** Let $\{\chi_j\}_{j=1}^{h(H)}$ be the set of all distinct irreducible characters of H. Then the number r of linearly independent elements in the induced characters $\{\chi_j\uparrow_H^G\}$ of G is the same as the number s of conjugacy classes C of G which meet H. (Hints: Pick $g_i \in C_i \cap H$, $1 \le i \le s$. The rank of the matrix $A = \left(\chi_i\uparrow_H^G(g_j)\right)$ is r and that of $(^tA)B$ is s where $B = (\chi_i(g_j^{-1}))$.)

10. Let A be an *abelian normal* subgroup of G. Show that the dimension of any irreducible representation of G divides the index $[G : A]$ of A in G. (Hint: Induction on $|G|$ using (4.4.3) and (3.10.7) above.)

11. Let H and N be two groups such that H acts on N as a group of automorphisms of N. Let $G = N \bullet H$ be the *semi-direct product* of H by N, i.e., as a set $G = N \times H$ and the group multiplication is given by $(a, x)(b, y) = (a(x \cdot b), xy)$ for all $a, b \in N$ and $x, y \in H$. Prove the following. (i) The subgroups $N \times \{1\}$ and $\{1\} \times H$ of G are respectively isomorphic to N and H with N *normal* and every element $z \in G$ can be written *uniquely* as $z = ax$ for $a \in N$ and $x \in H$. (ii) H is *also normal* in G $\iff$ H acts trivially on N (i.e., $x \cdot a = a$, $\forall x$ and a) $\iff$ G is the *direct product* of N and H. (iii) The semi-direct product $(\mathbf{Z}/3\mathbf{Z}) \bullet (\mathbf{Z}/2\mathbf{Z})$ is the non-abelian group of order 6 for the natural action of $\mathbf{Z}/2\mathbf{Z} = \mathrm{Aut}_\mathbf{Z}(\mathbf{Z}/3\mathbf{Z})$ on $\mathbf{Z}/3\mathbf{Z}$ as the full group of automorphisms of $\mathbf{Z}/3\mathbf{Z}$.

12. Show that the dihedral group $\mathbf{D}_{2\ell}$, of order 2ℓ (Ex.(3.11.10) above), is isomorphic to the semi-direct product $(\mathbf{Z}/\ell\mathbf{Z}) \bullet (\mathbf{Z}/2\mathbf{Z})$ (of $\mathbf{Z}/2\mathbf{Z}$ by $\mathbf{Z}/\ell\mathbf{Z}$) for the natural action of $\mathbf{Z}/2\mathbf{Z}$ on any cyclic group $< a >$, sending $a \mapsto a^{-1}$. Use this to give another proof that $\mathbf{D}_8 \not\cong Q_8$.

13. **(Mackey):** Show that the non-trivial character of the centre of Q_8 is *not* the restriction of any character of Q_8. (Does this contradict (4.7.9) above?)

14. Let $(\mathbf{H}_\mathbf{R}, \theta)$ be the representation of Q_8 on $\mathbf{H}_\mathbf{R}$ (over $\mathbf{R}$) by left multiplication. Let $\bar\theta$ be the representation of Q_8 on $\mathbf{H}_\mathbf{R} \otimes_\mathbf{R} \mathbf{C}$ (over $\mathbf{C}$) by the base change. Show that θ is irreducible over $\mathbf{R}$ whereas $\bar\theta$ is not irreducible but isotypical over $\mathbf{C}$. (Use Ex.(2.9.13) above.)

15. Let $\ell = 2m + 1 \geq 3$ be *odd* in which case write $\mathbf{D}^- = \mathbf{D}_{2\ell}$. Let the conjugacy classes of $\mathbf{D}^-$ (which are $m + 2$ in number) be labelled as in Ex.(3.11.30) above. Show that a complete set of inequivalent irreducible representations of $\mathbf{D}^-$ is given by $\mathrm{Irr}_K(\mathbf{D}^-) = \{(V_k, \varrho_k) \mid 1 \leq k \leq m + 2\}$ where

k	V_k	$\varrho_k(a)$	$\varrho_k(b)$
1	K	1	1
2	K	1	-1
$\begin{cases} 2+j \\ (1 \leq j \leq m) \end{cases}$	K^2	$\begin{pmatrix} \zeta^j & 0 \\ 0 & \zeta^{-j} \end{pmatrix}$	$\begin{pmatrix} 0 & 1 \\ 1 & 0 \end{pmatrix}$

(ζ is a primitive ℓ^{th} root of 1). Now write down the character table.

16. Let $\ell = 2m \geq 4$ be *even* in which case write $\mathbf{D}^+ := \mathbf{D}_{2\ell}$. Let the conjugacy classes of $\mathbf{D}^+$ (which are $m + 3$ in number) be labelled as in Ex.(3.11.29) above. Describe the set $\mathrm{Irr}_K(\mathbf{D}^+)$ and verify that
(i) the 2–dimensional representations are $m - 1$ in number, in fact, they are the same as the ones described above for $\mathbb{D}^-$ corresponding to $k = 2 + j$ with $1 \leq j \leq m - 1$ and
(ii) the character table is given as follows.

	C_0	$\begin{cases} C_k \\ (1 \leq k \leq m) \end{cases}$	C_b	C_{ab}
χ_1	1	1	1	1
χ_2	1	1	-1	-1
χ_3	1	$(-1)^k$	1	-1
χ_4	1	$(-1)^k$	-1	1
$\begin{cases} \chi_{4+j} \\ (1 \leq j \leq m-1) \end{cases}$	2	$\zeta^{jk} + \zeta^{-jk}$	0	0

(ζ is a primitive ℓ^{th} root of 1.) ∎

4.9 True/False Statements

Determine which of the following statements are true (T) or false (F) or partially true (PT). Justify your answers by giving a proof if (T) or providing a counter–example if (F)/(PT) or supplying the additional hypothesis needed to make it (T) (along with a proof) if (PT), as the case may be.

In what follows, G stands for a finite group and K an algebraically closed field such that $K[G]$ is semi-simple. All representations considered are assumed to be over K and finite dimensional.

1. Induced representation of a regular representation is regular.

2. Induced representation of a trivial representation is trivial.

3. Induced representation of an irreducible representation is one such.

4. Induced representation of an irreducible representation from a normal subgroup is irreducible.

5. Induced representation of an irreducible representation from H is irreducible if H is its own normaliser.

6. Induced representation of an irreducible representation is isotypical.

7. Induced representation of an irreducible representation is multiplicity free.

8. Restriction and induction commute with each other.

9. Restriction of an irreducible representation is one such.

10. Restriction of an irreducible representation to a normal subgroup is isotypical.

11. Restriction of an irreducible representation to a central subgroup is isotypical.

12. Restriction of an irreducible representation to a central subgroup is multiplicity free.

13. Restriction of the regular representation is regular.

14. Restriction of a trivial representation is trivial.

15. Multiplicity of W in $\operatorname{Res}\left(\operatorname{Ind}\left(W\right)\!\uparrow_H^G\right)\!\downarrow_H^G$ is positive.

16. Multiplicity of V in $\operatorname{Ind}\left(\operatorname{Res}\left(V\right)\!\downarrow_H^G\right)\!\uparrow_H^G$ is positive.

17. The character $\chi\!\uparrow_H^G$ is an integral multiple of χ on H if H is normal.

18. The character $\chi\!\uparrow_H^G$ vanishes outside H if H is normal.

19. Inertia group of a representation of a central subgroup H is G.

20. Inertia group I_W of a representation W of a subgroup $H \subseteq G$ is the normaliser of H in G. ∎

Part III

Representations of the SYMMETRIC AND ALTERNATING GROUPS

Chapter 5

Representations of the
Symmetric Group S_n

In this chapter, we shall determine all (ordinary) irreducible representations of the symmetric group S_n of permutations on n symbols, i.e.,
representations over an algebraically closed field K of characteristic
0. Since most (but not all) of the results go through even if the characteristic of K is non–zero but such that $K[S_n]$ is semi–simple, i.e.,
Char $K > n$, we shall allow K to have positive characteristic and
specify characteristic 0 wherever necessary. As remarked in (3.7.6)
above, we need to do the following four things.

1. Determine the conjugacy classes of S_n.

2. For each conjugacy class λ, construct an irreducible representation
V_λ in such a way that

3. V_λ is *not equivalent* to V_μ for $\lambda \neq \mu$ and finally

4. determine the dimensions of the V_λ's.

We shall present FOUR METHODS: one for determining the characters (without actually constructing the representation spaces) and the
other three for constructing the V_λ's, namely,

I. FROBENIUS method of determining the irreducible characters.

II. FROBENIUS–YOUNG method of realising the irreducibles as the
minimal left ideals in $K[S_n]$.

III. SPECHT method of explicitly constructing simple

$K[S_n]$–modules, as suitable subspaces of the polynomial algebra in n variables over K.

IV. An ABSTRACT method using the decomposition of certain induced representations.

Determination of the dimensions, leading to the so called HOOK–LENGTH FORMULA (5.8.5), is the *non–trivial* part. (A full proof of this formula is not given but an outline is sketched.)

Having at hand the required information for the Symmetric groups S_n, we shall see that the task becomes easy or even formal for the other groups B_n, A_n and D_n. For this reason, we present the theory for S_n keeping the technicalities to a minimum. For example, the methods I, II and III do not require any of the considerations of Chapter 4. (As presented here for simplicity, the method IV uses some facts from the method II which can be avoided.) (Cf. [19], [24], [25], [32], [35], [48] and several other papaers and books cited in the bibliography.)

5.1 The Symmetric Group S_n

Let us recall some of the basic facts about the Symmetric group S_n and set the notation. We use the standard notation for a permutation θ on the symbols $1, \cdots, n$, namely,

$$\theta = \begin{pmatrix} 1 & 2 & \cdots & i & \cdots & n \\ \theta(1) & \theta(2) & \cdots & \theta(i) & \cdots & \theta(n) \end{pmatrix}.$$

A permutation θ, treated as a bijective map of the set $\{1, \cdots, n\}$ onto itself, is simply the map $i \mapsto \theta(i)$. In conformity with the composition rule for maps, we shall multiply permutations from *right to left*; for instance, we have

$$\begin{pmatrix} 1 & 2 & 3 & 4 & 5 \\ 3 & 1 & 4 & 5 & 2 \end{pmatrix} \cdot \begin{pmatrix} 1 & 2 & 3 & 4 & 5 \\ 5 & 3 & 4 & 2 & 1 \end{pmatrix} = \begin{pmatrix} 1 & 2 & 3 & 4 & 5 \\ 2 & 4 & 5 & 1 & 3 \end{pmatrix}.$$

5.1.0 The set S_n of all permutations on n symbols is a group of order $n!$ under composition of permutations.

5.1.1 By a *cycle* $\theta = (a_1, \cdots, a_r)$, we mean the permutation which maps $a_1 \mapsto a_2, a_2 \mapsto a_3, \cdots, a_{r-1} \mapsto a$ and $a_r \mapsto a_1$ and all other symbols to themselves. The number r is called the *length of the cycle* and we call θ an r–cycle. The order of an r–cycle in the group S_n is simply r, i.e., $\theta^r = 1$ and $\theta^s \neq 1$ if $s < r$. Any 1–cycle is called a *trivial cycle* and it is simply the identity permutation.

5.1.2 A conjugate of an r–cycle is again an r–cycle. In fact, if $\theta = (a_1, \cdots, a_r)$ is an r–cycle and $\sigma \in S_n$, we have
$$\sigma\theta\sigma^{-1} = \big(\sigma(a_1), \cdots, \sigma(a_r)\big).$$

5.1.3 Two cycles $(a_1, \cdots, a_r)$ and $(b_1, \cdots, b_s)$ are said to be *disjoint* if $\{a_1, \cdots, a_r\} \cap \{b_1, \cdots, b_s\} = \emptyset$. It follows trivially that disjoint cycles commute with each other.

5.1.4 By a *transposition* we mean a 2–cycle.

5.1.5 Every permutation θ can be written as a product of transpositions $\theta = \prod_{i=1}^{m} \theta_i$ where $\theta_i = (a_i, b_i)$. Eventhough the transpositions θ_i or the number m of them in the above product are not unique, the *parity* of the number of transpositions is the same in any decomposition into product of transpositions, i.e., $(-1)^m$ is independent of the decomposition chosen and is called the *signature* or the *sign* of θ and is denoted by $\mathrm{sgn}(\theta)$. It is clear that 'sgn' is multiplicative.

5.1.6 A permutation is called *even* or *odd* according as it can be written as a product of even or odd number of transpositions, i.e., according as its sign is $+1$ or -1.

An r–cycle $\theta = (a_1, \cdots, a_r)$ is a product of $r - 1$ transpositions, namely, $\theta = (a_1, \cdots, a_r) = \prod_{i=1}^{r-1}(a_i, a_{i+1})$, and hence an r–cycle is even or odd according as r is odd or even.

5.1.7 The set of all even permutations is a subgroup of index 2 in S_n, called the *alternating group* on n symbols and is denoted by A_n.

5.1.8 Every permutation can be uniquely written as a product of disjoint cycles, including the trivial cycles, say $\theta = \prod_{i=1}^{r} \theta_i$ where θ_i is a λ_i–cycle, $1 \leq i \leq r$. Since these cycles commute with each other, we can assume that $\lambda_1 \geq \cdots \geq \lambda_r \geq 1$.

5.1.9   The   decreasing   sequence   of   positive   integers $\lambda =$

$(\lambda_1, \cdots, \lambda_r)$ in the decomposition of θ above is called the *cycle type* of θ. We have $\sum_{i=1}^{r} \lambda_i = n$.

For example, we have

$$\begin{pmatrix} 1 & 2 & 3 & 4 & 5 \\ 3 & 1 & 4 & 5 & 2 \end{pmatrix} = (1,3,4,5,2),$$

$$\begin{pmatrix} 1 & 2 & 3 & 4 & 5 \\ 5 & 3 & 4 & 2 & 1 \end{pmatrix} = (2,3,4)(1,5) \text{ and}$$

$$\begin{pmatrix} 1 & 2 & 3 & 4 & 5 \\ 3 & 5 & 1 & 4 & 2 \end{pmatrix} = (1,3)(2,5)(4).$$

and hence their cycle types are respectively (5), (3,2) and (2,2,1).

5.1.10   The cycle type of a permutation is invariant under conjugation. (This follows from (5.1.2) and (5.1.8) because if $\theta, \sigma \in S_n$, we have $\sigma\theta\sigma^{-1} = \prod_{i=1}^{r} \sigma\theta_i\sigma^{-1}$ where θ_i is a λ_i-cycle, $1 \le i \le r$.)

5.1.11 The commutator subgroup of S_n is A_n. In fact, more is true, namely, A_n is the set of all commutators in S_n. (It is rare that the set of all commutators is a subgroup.) This can be seen as follows.

Given an even permutation θ, we may assume that the cycle decomposition of θ is arranged in the form $\theta = (\theta_1 \cdots \theta_r)(\vartheta_1\phi_1) \cdots (\vartheta_s\phi_s)$, where the θ_j are cycles of odd length and ϑ_k, ϕ_k are pairs of cycles of even (but perhaps of unequal) lengths. If we can express each of the θ_j's (resp. the products $\vartheta_k\phi_k$) as a commutator involving only the symbols of θ_j (resp. $\vartheta_k\phi_k$), it would follow that θ is a commutator since the factors of any of these commutators commutes with those of any other, being based on disjoint sets of symbols.

Since a conjugate of a commutator is again a commutator, we may assume that the cycles of θ are on subsets of consecutive integers. If $\rho = (1, 2, \cdots, 2\ell + 1)$, then we have

$$\rho = (1, 2, \cdots, \ell + 1)(\ell + 1, \cdots, 2\ell + 1) = \sigma\tau\sigma^{-1}\tau^{-1} \text{ where}$$

$$\sigma = (1, 2, \cdots, \ell + 1) \text{ and } \tau = \begin{pmatrix} 1 & 2 & \cdots & \ell & \ell + 1 \\ 2\ell + 1 & 2\ell & \cdots & \ell + 2 & \ell + 1 \end{pmatrix}.$$

Similarly, if

$$\varrho = (1, \cdots, 2\ell)(2\ell, \ell + m + 1, \cdots, 2\ell + 2m) \text{ with } \ell \le m,$$

then we have $\varrho = \xi\psi\xi^{-1}\psi^{-1}$ where

$$\xi = (1, \cdots, \ell + m + 1) \text{ and } \psi = \begin{pmatrix} 1 & \cdots & \ell + m + 1 \\ 2\ell + 2m & \cdots & 2\ell \end{pmatrix}.$$

5.1.12 The group of characters $\widetilde{S_n}$ of S_n is of order 2. In fact, we have $\widetilde{S_n} = \{\iota, \varepsilon\}$ where ι is the *trivial character* and ε is the *sign character*, i.e., $\iota(\theta) = 1$ and $\varepsilon(\theta) = \mathrm{sgn}(\theta)$ for all $\theta \in S_n$.   ∎

5.2   The Conjugacy Classes of S_n

5.2.1 **Partition of n:** A decreasing sequence of positive integers $\lambda = (\lambda_1, \cdots, \lambda_r)$, i.e., $\lambda_1 \geq \cdots \geq \lambda_r \geq 1$, is called a *partition* of n if $\sum_{i=1}^r \lambda_i = n$ and it is denoted by $\lambda \vdash n$. The λ_i's are called the *parts* and the integer r is called the *number of parts* or the *depth* of λ.

Two partitions $\lambda = (\lambda_1, \cdots, \lambda_r)$ and $\mu = (\mu_1, \cdots, \mu_s)$ of n are said to be *equal* if $r = s$ and $\lambda_i = \mu_i$ for all i.

The set of all partitions of n is denoted by $P(n)$ and the number of partitions of n is denoted by $p(n)$.

Notation: Given a $\lambda \vdash n$, we write $\lambda = (n^{\gamma_n}, \cdots, i^{\gamma_i}, \cdots, 2^{\gamma_2}, 1^{\gamma_1})$ where γ_i is the number of times i occurs as a part in λ. We have $\gamma_i \in \mathbf{Z}^+$ and $r = \sum_{i=1}^n \gamma_i$ is the number of parts of λ.

5.2.2 **Dictionary order:** On the set of partitions of n, we have a natural total order $\succeq$, called the *dictionary order*, namely, for $\lambda = (\lambda_1, \cdots, \lambda_r)$ and $\mu = (\mu_1, \cdots, \mu_s)$ in $P(n)$, we say that $\lambda \succeq \mu$, if $\lambda = \mu$ or the first non–zero element in the sequence of differences $\{\lambda_i - \mu_i\}_{i \geq 1}$ is positive. (The largest and least elements of $P(n)$ are (n) and (1^n) respectively.)

5.2.3 **Strictly decreasing partition (s.d.p):** A partition $\lambda = (\lambda_1, \cdots, \lambda_r)$ is called a *strictly decreasing partition*, abreviated as s.d.p., if $\lambda_1 > \lambda_2 > \cdots > \lambda_r$, i.e., all the parts of λ are distinct. Or, writing $\lambda = (n^{\gamma_n}, \cdots, i^{\gamma_i}, \cdots, 2^{\gamma_2}, 1^{\gamma_1})$, we have $\gamma_i \leq 1$, $\forall i$.

5.2.4 **Examples:** 1. $P(1) = \{(1)\}$ and $p(1) = 1$.
2. $P(2) = \{(2), (1^2)\}$, arranged in the descending order and

$p(2) = 2$. Only one is strictly decreasing.

3. $P(5) = \{(5),(4,1),(3,2),(3,1^2),(2^2,1),(2,1^3),(1^4)\}$, arranged in the descending order and $p(5) = 7$. Only the first three are s.d.p.

4. The cycle type of any permutation in S_n is a partition of n and conversely, given any partition $\lambda \vdash n$, there is a $\theta \in S_n$ whose cycle type is λ. For example, if $\lambda = (\lambda_1, \lambda_2, \cdots, \lambda_r)$, let θ_i be the λ_i–cycle, namely, $\theta_i = (a_{i-1}+1, a_{i-1}+2, \cdots, a_{i-1}+\lambda_i)$ where $a_{i-1} = (\sum_{j=0}^{i-1}\lambda_j)$ with $\lambda_0 = 0$. Now take $\theta = \prod_{i=1}^{r}\theta_i$.

5.2.5 Theorem: *The set of all conjugacy classes of S_n is naturally bijective with the set of all partitions of n.*

Proof: Since the cycle type of a permutation in S_n is a partition of n, it suffices to prove that two permutations are conjugates of each other if and only if their cycle types are equal. Let $\theta, \vartheta \in S_n$. If θ and ϑ are conjugates, then by (5.1.11) above, they have the same cycle type. Conversely, suppose they have the same cycle type, say $\lambda = (\lambda_1, \cdots, \lambda_r) \vdash n$. Let the cycle decompositions of θ and ϑ be $\theta = \prod_{i=1}^{r}\theta_i$ and $\vartheta = \prod_{i=1}^{r}\vartheta_i$. Define

$$\sigma = \begin{pmatrix} \theta_1 & \theta_2 & \cdots & \theta_i & \cdots & \theta_r \\ \vartheta_1 & \vartheta_2 & \cdots & \vartheta_i & \cdots & \vartheta_r \end{pmatrix}$$

where σ takes θ_i to ϑ_i elementwise. Now we have

$$\sigma\theta\sigma^{-1} = \prod_{i=1}^{r}\sigma\theta_i\sigma^{-1} = \prod_{i=1}^{r}\vartheta_i = \vartheta, \text{ as required.} \qquad \Diamond$$

5.2.6 Proposition: *Let $\lambda = (n^{\gamma_n}, \cdots, i^{\gamma_i}, \cdots, 2^{\gamma_2}, 1^{\gamma_1}) \vdash n$. Let C_λ be the conjugacy class of cycle type λ and $|C_\lambda|$ be its cardinality. Then we have*

$$|C_\lambda| = \frac{n!}{1^{\gamma_1}\gamma_1! \cdot 2^{\gamma_2}\gamma_2! \cdots n^{\gamma_n}\gamma_n!}.$$

In particular, the number of j–cycles in S_n is given by

$$\left|C_{(j,1^{n-j})}\right| = \frac{n!}{1^{n-j}(n-j)!j} = \frac{n!}{(n-j)!j} = \binom{n}{j}(j-1)!.$$

Proof: The elements of C_λ are obtained by filling a system of γ_j blank j–cycles, $1 \leq j \leq n$, with the entries $1, 2, \cdots, n$ in all the $n!$

possible ways without repetitions. Since cyclic permutation of entries in any cycle does not change the cycle and disjoint cycles commute, we see that among the $n!$ different fillings which give all the elements of S_n, each element $x \in C_\lambda$ appears exactly ℓ_x times where $\ell_x = 1^{\gamma_1}\gamma_1!\cdot 2^{\gamma_2}\gamma_2!\cdots n^{\gamma_n}\gamma_n!$, i.e., the product of all the j-cycles of x repeats $j^{\gamma_j}\gamma_j!$ times in the fillings.   ■

5.3   I. Irreducible Characters of S_n

In this section, we shall briefly mention Frobenius' method of determining the complete set of (ordinary) irreducible characters of S_n without actually constructing the corresponding representation spaces. Since we do not use these results in the sequel, we omit the details. The reader may refer to the exposition in [32], pp.94–111.

Recall that the (ordinary) representations of any finite group are completely determined by their characters. The values of the characters at the identity element of the group give the dimensions of the corresponding representation spaces. Frobenius' method rests on first knowing the permutation characters (3.4.4) of S_n defined by a class of "Young subgroups", one for each partition of n and then showing that this set is a basis for the additive group of all characters of S_n. Thereafter, it is not difficult to find the irreducible ones and their dimensions, etc. All this needs a chasing of a rather formidable bunch of polynomial and determinantal identities and so on.

5.3.1   Young subgroups: Given a partition $\lambda = (\lambda_1, \cdots, \lambda_r) \vdash n$, a subgroup of S_n of the form $Y_\lambda = S_{\lambda_1} \times \cdots \times S_{\lambda_r}$ is called a *Young subgroup* of S_n of shape λ.

The order of a Young subgroup is $\lambda_1! \cdots \lambda_r!$ and hence its index is $n!/(\lambda_1! \cdots \lambda_r!)$. We shall now determine the permutation character defined by a Young subgroup of shape λ. (Note that such a character is integer valued.)

5.3.2   Proposition (Permutation characters):   *For $\lambda \vdash n$, let $(K^{[S_n:Y_\lambda]}, \mathrm{perm}_{Y_\lambda})$ be the permutation representation of S_n defined by* *the Young subgroup Y_λ. Then the value of its character $\zeta_\lambda = \chi_{\text{perm}_{Y_\lambda}}$ on the conjugacy class C_μ is given by*

$$(\star): \qquad \zeta_\lambda(C_\mu) = [S_n : Y_\lambda]\frac{|C_\mu \cap Y_\lambda|}{|C_\mu|} = \frac{n!\,|C_\mu \cap Y_\lambda|}{(\lambda_1! \cdots \lambda_r!)\,|C_\mu|}.$$

Proof: This is a special case of (3.4.4) above. $\Diamond$

The permutation character ζ_λ is completely known once we know $|C_\mu \cap Y_\lambda|$ since $|C_\mu|$ is known for all $\mu \vdash n$ by (5:2.6) above. Let us now calculate $|C_\mu \cap Y_\lambda|$. Let us write
$\lambda = (\lambda_1, \cdots, \lambda_r) = (n^{\gamma_n}, \cdots, j^{\gamma_j}, \cdots, 1^{\gamma_1})$ and
$\mu = (\mu_1, \cdots, \mu_s) = (n^{\varepsilon_n}, \cdots, j^{\varepsilon_j}, \cdots, 1^{\varepsilon_1})$.
An element $\theta \in Y_\lambda$ is of the form $\theta = \theta_1 \cdots \theta_r$ for unique $\theta_j \in S_{\lambda_j}$, $1 \le j \le r$, where S_{λ_j} is the subgroup of S_n on the λ_j successive symbols $\{(\sum_{k=0}^{j-1}\lambda_k) + 1, \cdots, \sum_{k=0}^{j}\lambda_k\}$ with $\lambda_0 = 0$.

Let ν_{jk} be the number of k–cycles in the cycle decomposition of θ_j, $1 \le k \le n$, so that we have

$$(A): \qquad \lambda_j = \sum_{k=1}^{n} k\nu_{jk}, \quad 1 \le j \le r.$$

If $\theta \in C_\mu$, then the total number of k–cycles in θ must add up to ε_k for all k. This gives another relation

$$(B): \qquad \varepsilon_k = \sum_{j=1}^{r} \nu_{jk}, \quad 1 \le k \le n.$$

Thus every element $\theta \in C_\mu \cap Y_\lambda$ gives rise to an $r \times n$ matrix (ν_{jk}) of non–negative integers satifying the equations (A) and (B) above and conversely. We shall now find the number of elements $\theta \in C_\mu \cap Y_\lambda$ corresponding to the same matrix (ν_{jk}) satisfying (A) and (B).

For a fixed j, the equation (A) specifies a conjugacy class of S_{λ_j} of cycle type $(n^{\nu_{jn}}, \cdots, k^{\nu_{jk}}, \cdots, 1^{\nu_{j1}})$ which is a partition of λ_j and hence by (5.2.6) above, it contains

$$h^{(j)}\big((\nu_{jk})\big) = \frac{\lambda_j!}{1^{\nu_{j1}}\nu_{j1}! \cdots n^{\nu_{jn}}\nu_{jn}!}$$

elements. Now letting j range from 1 to r, we obtain that $C_\mu \cap Y_\lambda$ contains $\prod_{j=1}^r h^{(j)}\big((\nu_{jk})\big)$ elements corresponding to the same matrix (ν_{jk}) satisfying the equations (A) and (B). Finally letting the matrices vary over all possible solutions of (A) and (B), we get that

$$\left| C_\mu \cap Y_\lambda \right| = \sum_{A,B} \prod_{j=1}^r h^{(j)}\big((\nu_{jk})\big) = \sum_{A,B} \prod_{j=1}^r \frac{\lambda_j!}{1^{\nu_{j1}}\,\nu_{j1}!\cdots n^{\nu_{jn}}\,\nu_{jn}!}.$$

Substituting the values of $\left| C_\mu \cap Y_\lambda \right|$ and $\left| C_\mu \right|$ in $(\star)$ above, and on straightforward simplication, we have proved the following.

5.3.3 Theorem: *The value of the permutation character ζ_λ defined by a Young subgroup of shape λ on a conjugacy class of cycle type μ is given by*

$$\zeta_\lambda(C_\mu) = \sum_{A,B} \prod_{k=1}^n \frac{\varepsilon_k!}{\nu_{k1}!\cdots\nu_{kn}!}.$$

This is an integer valued class function on S_n. $\diamond$

Let $\{\xi_\lambda \mid \lambda \vdash n\}$ be the set of all irreducible characters of S_n. We know that this is an integral basis of the additive group of all characters of S_n. Using the precise values of the characters ζ_λ, as above, and chasing a host of polynomial and determinantal identities, one gets the following.

5.3.4 Theorem (Frobenius): *The set $\{\zeta_\lambda\}_{\lambda \vdash n}$ of permutation characters defined by the respective Young subgroups is another basis for the group of characters of S_n. In particular, all characters of S_n are integer valued.*

Proof: See [32] pp. 94–111. See also (5.7.12) below, for another proof of the last assertion. $\diamond$

5.3.5 Remark: Writing $\xi_\lambda = \sum_{\mu \vdash n} a_{\mu,\lambda}\zeta_\mu$ (where $a_{\mu,\lambda} \in \mathbb{Z}$) and estimating the coefficients $a_{\mu,\lambda}$'s in terms of the character values $\zeta_\mu(C_\lambda)$, as in (5.3.3) above, the degrees of the irreducible characters ξ_λ's are also seen to yield the "hook–length formula", as done in [32] and explained in §5.8, below. ∎

5.4   Young Diagrams and Tableaux

5.4.1 Young diagram: Given a partition $\lambda = (\lambda_1, \cdots, \lambda_r) \vdash n$, by a *Young diagram T_λ of shape* λ, we mean a left and top aligned *frame* of empty boxes having r rows and λ_1 columns such that the ith–row has λ_i boxes for $1 \leq i \leq r$. It is the following diagram.

$$T_\lambda \, :$$

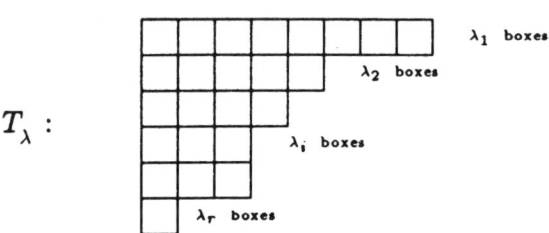

The above is a Young diagram of shape $\lambda = (8, 5, 4, 3^2, 1) \vdash 24$.

5.4.2 Conjugate partitions: Given a partition $\lambda = (\lambda_1, \cdots, \lambda_r) \vdash n$, the partition $\lambda' = (\lambda'_1, \cdots, \lambda'_s) \vdash n$ is called the *conjugate* of λ where λ'_i is the number of boxes in the ith column of the Young diagram T_λ of shape λ.

We note that the Young diagram $T_{\lambda'}$ of shape λ' is the transpose of T_λ and hence $s = \lambda_1$, $\lambda'_1 = r$ and $\lambda'' = \lambda$. The Young diagram of shape λ' is the following.

$$T_{\lambda'} \, :$$

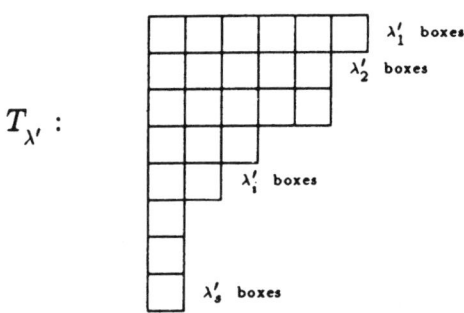

The above is a Young diagram of shape $\lambda' = (6, 5^2, 3, 2, 1^3) \vdash 24$ where $\lambda = (8, 5, 4, 3^2, 1)$.

5.4.3 Young tableaux: Given a partition $\lambda = (\lambda_1, \cdots, \lambda_r) \vdash n$, by a *Young tableaux*, we mean a Young diagram T_λ of *shape* λ whose boxes are filled with the integers between 1 and n without repetition.

It is clear that there are exactly $n!$ Young tableaux of any given shape λ because we can parametrise them by the elements of S_n, as follows.

Convention: Given $\lambda \vdash n$ and $\sigma \in S_n$, the Young tableaux $T_\lambda(\sigma)$ of shape λ *associated to* or *filled along* σ is the diagram T_λ which is filled with the entries $\sigma(1), \cdots, \sigma(n)$ *down the columns starting with the first, left to right*. For example, we have the following.

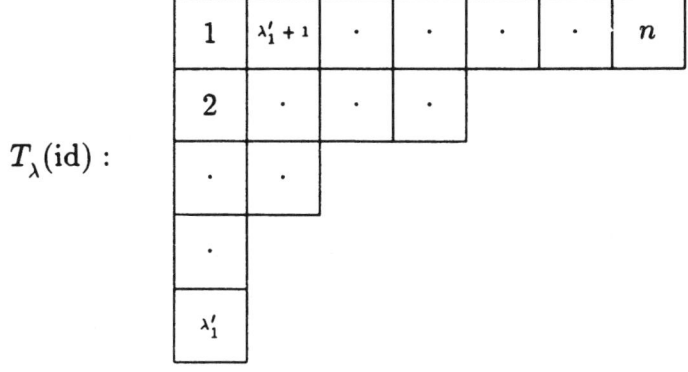

In particular, if $\sigma = \text{id}$, then we have the following.

$$T_\lambda(\text{id}):$$

The group S_n acts on the set of Young tableaux of any given shape by permuting the entries of the tableaux, i.e., we have $\sigma(T_\lambda(\theta)) = T_\lambda(\sigma\theta)$ for all σ, $\theta \in S_n$ and $\lambda \vdash n$. We have $\sigma(T_\lambda(\mathrm{id})) = T_\lambda(\sigma)$.

5.4.4 Row group: Given a Young tableaux $T_\lambda(\theta)$ (of a given shape λ), the set of all permutations which keep the rows stable (but not necessarily entrywise) is a subgroup of S_n, called the *row group* of the tableaux and is denoted by $R(T_\lambda(\theta))$ or simply R_λ when there is no confusion likely.

5.4.5 Column group: The *column group* of a tableaux is the subgroup of S_n which keeps the columns stable and is denoted by C_λ.

5.4.6 Proposition: *If* $\lambda = (\lambda_1, \cdots, \lambda_r) \vdash n$ *and* T_λ *is a fixed Young tableaux of shape* λ, *then we have the following.*
1. $C(T_\lambda) = R(T_{\lambda'})$, *i.e., the column group of a tableaux is the row group of its transpose.*
2. $R(T_\lambda) \simeq \prod_{i=1}^r S_{\lambda_i}$ *where* S_{λ_i} *is the permutation group on the entries of the* i^{th} *row of the tableaux,* $1 \leq i \leq r$, *i.e.,* $R(T_\lambda)$ *is a Young subgroup of shape* λ (5.3.1).
3. $R(T_\lambda) \cap C(T_\lambda) = \{1\}$ *and*
4. $R(\sigma T_\lambda) = \sigma R(T_\lambda)\sigma^{-1}$ *and* $C(\sigma T_\lambda) = \sigma C(T_\lambda)\sigma^{-1}$ *for all* $\sigma \in S_n$.

Proof: Easy verification.   ∎

5.5   II. Frobenius–Young Modules for S_n

Given a partition $\lambda \vdash n$, we now define a minimal left ideal V_λ of $K[S_n]$ and show that the family V_λ, $\lambda \vdash n$ gives a complete set of mutually non–isomorphic simple $K[S_n]$–modules, called the *Frobenius-Young modules* of $K[S_n]$.

For $\lambda \vdash n$, we fix a Young tableaux T_λ of shape λ. Let the row and column groups of T_λ be R_λ and C_λ respectively. Let

$$a_\lambda = \sum_{\sigma \in R_\lambda} \sigma, \quad b_\lambda = \sum_{\tau \in C_\lambda} \mathrm{sgn}(\tau)\tau \quad \text{and} \quad c_\lambda = a_\lambda b_\lambda.$$

5.5.0 Examples: 1. Let $\lambda = (n)$. Then $T_{(n)}$ is a single row consisting of n boxes filled with a permutation in S_n. It is clear that $R_{(n)} = S_n$ and $C_{(n)} = \{1\}$ and hence $b_{(n)} = 1$ and $a_{(n)} = c_{(n)} = \sum_{\theta \in S_n} \theta$ which is a central element of $K[S_n]$. In fact, we have $\theta\, c_{(n)} = c_{(n)}\theta = c_{(n)}$ for all $\theta \in S_n$.

2. Let $\lambda = (1^n)$. Then $T_{(1^n)}$ is a single column consisting of n boxes filled with a permutation in S_n. It is clear that $R_{(1^n)} = \{1\}$ and $C_{(1^n)} = S_n$ and hence $a_{(1^n)} = 1$ and $b_{(1^n)} = \sum_{\theta \in S_n} \mathrm{sgn}(\theta)\theta$ which is again a central element of $K[S_n]$. In fact, we have $\theta\, c_{(1^n)} = c_{(1^n)}\theta = \mathrm{sgn}(\theta)c_{(1^n)}$ for all $\theta \in S_n$.

5.5.1 Proposition: *With notation as above, we have the following.*
1. $c_\lambda \neq 0$,
2. $\sigma a_\lambda = a_\lambda \sigma = a_\lambda$ *for all* $\sigma \in R_\lambda$,
3. $\tau b_\lambda = b_\lambda \tau = \mathrm{sgn}(\tau)b_\lambda$ *for all* $\tau \in C_\lambda$,
4. $\sigma c_\lambda = c_\lambda$ *for all* $\sigma \in R_\lambda$ *and*
5. $c_\lambda \tau = \mathrm{sgn}(\tau)c_\lambda$ *for all* $\tau \in C_\lambda$.

Proof: By definition, we have
$$c_\lambda = \Big(\sum_{\sigma \in R_\lambda} \sigma\Big)\Big(\sum_{\tau \in C_\lambda} \mathrm{sgn}(\tau)\tau\Big) = \sum_{\sigma \in R_\lambda}\sum_{\tau \in C_\lambda} \mathrm{sgn}(\tau)\sigma\tau.$$
Since $R_\lambda \cap C_\lambda = \{1\}$, it follows that $\sigma\tau \neq \sigma'\tau'$ for $(\sigma,\tau), (\sigma',\tau')$ in $R_\lambda \times C_\lambda$ with $(\sigma,\tau) \neq (\sigma',\tau')$. Hence, c_λ is a sum (in $K[S_n]$) of some distinct elements of S_n and so $c_\lambda \neq 0$. This proves **(1)**. The others are immediate consequences of definitions and the fact that R_λ and C_λ are subgroups of S_n. ◇

5.5.2 Lemma (Von Neumann): *Let* $\lambda, \mu \vdash n$. *Suppose* $\lambda \succeq \mu$ *in the dictionary order of partitions. Let* T_λ *and* T_μ *be fixed Young tableaux of shapes* λ *and* μ *respectively. Then we have the following.*
1. *Either there exists a non-empty subset of* $\{1,\cdots,n\}$ *of cardinality at least 2 which occur in the same row of* T_λ *and in the same column of* T_μ, *or else,*
2. $\lambda = \mu$ *and* $T_\mu = \theta\tau T_\lambda$ *for some* $\theta \in R_\lambda$ *and* $\tau \in C_\lambda$.

Proof: Let $\lambda \succ \mu$ in the dictionary order of partitions. We may assume that $\lambda_1 > \mu_1$, i.e., the number of columns of T_λ is larger than that of T_μ. Hence atleast two entries of the first row of T_λ occur in the same column of T_μ, i.e., (1) holds if $\lambda \succ \mu$.

Now suppose (1) does not hold (hence $\lambda = \mu$), i.e., all the entries of any row of T_λ occur in different columns of T_μ. Applying a suitable column permutation τ_1 of T_μ, we can get that $\tau_1 T_\mu$ has its first row filled with the entries of the first row of T_λ. Similarly, applying a suitable row element θ_1 of T_λ, we can assume that the first row of $\theta_1 T_\lambda$ is identical with that of $\tau_1 T_\mu$. Repeating this argument with the second row of $\theta_1 T_\lambda$ keeping $\tau_1 T_\mu$ in place of T_μ, we get that the second rows of $\theta_2 \theta_1 T_\lambda$ and $\tau_2 \tau_1 T_\mu$ are identical for some $\theta_2 \in R_\lambda$ and $\tau_2 \in C_\mu$. Proceeding thus we get finally that $\theta T_\lambda = \tau T_\mu$ for some $\theta \in R_\lambda$ and $\tau \in C_\mu$. Now we have $T_\lambda = \theta^{-1} \tau T_\mu$, as required. $\diamond$

5.5.3 Corollary: *Suppose $\sigma \notin R_\lambda \cdot C_\lambda$, then there exist transpositions $u \in R_\lambda$ and $v \in C_\lambda$ such that $u\sigma = \sigma v$.*

For, take $T_\mu = \sigma T_\lambda$ and note that $\mu = \lambda$. Now the statement (2) of the above lemma is not true and hence by (1), there exists a pair (a, b) of integers between 1 and n such that a, b occur in the same row of T_λ and the same column of T_μ. Thus the transposition $u = (a, b) \in R_\lambda \cap C_\mu$. On the other hand, by (5.4.6)(4) above, we have $C_\mu = \sigma C_\lambda \sigma^{-1}$. Hence we get that $u = \sigma v \sigma^{-1}$ for some $v \in C_\lambda$. It is clear that $v = (\sigma^{-1}(a), \sigma^{-1}(b))$ is what we are looking for. $\diamond$

5.5.4 Lemma: *Let $x \in K[S_n]$. Then $x = ac_\lambda$ for some $a \in K$ if and only if $\sigma x \tau = sgn(\tau) x$ for all $\sigma \in R_\lambda$ and $\tau \in C_\lambda$.*

Proof: If $x = ac_\lambda$ for $a \in K$, the result follows trivially by (5.5.1) above. Conversely, suppose that $\sigma x \tau = \text{sgn}(\tau) x$ for all $\sigma \in R_\lambda$ and $\tau \in C_\lambda$. Let $x = \sum_{\theta \in S_n} a_\theta \theta$ with $a_\theta \in K$. Recall that we have
$c_\lambda = \sum_{\sigma \in R_\lambda} \sum_{\tau \in C_\lambda} \text{sgn}(\tau)\sigma\tau.$

Claim 1. We have $\sigma x \tau = \text{sgn}(\tau) x \Rightarrow a_{\sigma\tau} = \text{sgn}(\tau) a_1$.

For, we have $\sigma x \tau = \sum_{\theta \in S_n} a_\theta \sigma \theta \tau = \sum_{\vartheta \in S_n} a_{\sigma^{-1}\vartheta\tau^{-1}} \vartheta$ where $\vartheta = \sigma\theta\tau$. On the other hand, we have $\text{sgn}(\tau) x = \sum_{\theta \in S_n} a_\theta \, \text{sgn}(\tau)\theta$. Comparing the coefficient of ϑ on both sides, we get that $a_{\sigma^{-1}\vartheta\tau^{-1}} = \text{sgn}(\tau) a_\vartheta$ for all $\vartheta \in S_n$. In particular, for $\vartheta = 1$, we get that $a_{\sigma^{-1}\tau^{-1}} = \text{sgn}(\tau) a_1$ for all $\sigma \in R_\lambda$ and $\tau \in C_\lambda$, i.e., $a_{\sigma\tau} = \text{sgn}(\tau) a_1$, as required.

Claim 2. If $\theta \notin R_\lambda \cdot C_\lambda$, then $a_\theta = 0$.

For, by (5.5.3), there exist transpositions $u \in R_\lambda$ and $v \in C_\lambda$ such that $v = \sigma^{-1}u\sigma$, i.e., $u\sigma v = \sigma$, hence we get $a_{u\sigma v} = a_\sigma$. This implies (by hypothesis) that $\mathrm{sgn}(v)a_\sigma = a_\sigma$, i.e., $a_\sigma = -a_\sigma$ or $a_\sigma = 0$ since the characteristic of K is $\neq 2$. Finally we have

$$
\begin{aligned}
x &= \sum_{\theta \in S_n} a_\theta \theta = \sum_{\theta \in R_\lambda C_\lambda} a_\theta \theta + \sum_{\theta \notin R_\lambda C_\lambda} a_\theta \theta \\
&= \sum_{\sigma \in R_\lambda} \sum_{\tau \in C_\lambda} a_{\sigma\tau}\sigma\tau + 0 \quad \text{(by Claim 2)} \\
&= \sum_{\sigma \in R_\lambda} \sum_{\tau \in C_\lambda} a_1 \mathrm{sgn}(\tau)\sigma\tau \quad \text{(by Claim 1)} \\
&= a_1 c_\lambda, \quad \text{as required.} \qquad \diamond
\end{aligned}
$$

5.5.5 Theorem: *We have $c_\lambda^2 = ac_\lambda$ for some $a \in K$, $a \neq 0$.*

Proof: Taking $x = c_\lambda^2$ in the lemma above, we get that $c_\lambda^2 = ac_\lambda$ for some $a \in K$. We have only to show that $a \neq 0$. To this end, we proceed as follows.

Look at the right multiplication r_{c_λ} on $K[S_n]$ by c_λ, i.e., $x \mapsto xc_\lambda$. We know that r_{c_λ} is an S_n–linear homothecy. Let us calculate the trace of r_{c_λ} in two different ways: one with respect to the basis S_n of $K[S_n]$ and the other with respect to another basis to be specified.

1. To find the trace with respect to the basis S_n, we have to find the coefficient a_g of g in gc_λ for all $g \in S_n$ so that we would have $\mathrm{trace}(r_{c_\lambda}) = \sum_{g \in S_n} a_g$. We have

$$
\begin{aligned}
c_\lambda &= \sum_{\sigma \in R_\lambda} \sum_{\tau \in C_\lambda} \mathrm{sgn}(\tau)\sigma\tau = 1 + \sum_{\sigma\tau \neq 1} \mathrm{sgn}(\tau)\sigma\tau \quad \text{hence} \\
gc_\lambda &= g + \sum_{\sigma\tau \neq 1} \mathrm{sgn}(\tau)g\sigma\tau \quad \Rightarrow a_g = 1, \quad \forall g \in S_n \\
&\Rightarrow \mathrm{trace}(r_{c_\lambda}) = n! \neq 0 \text{ in } K \text{ (since Char } K > n \text{ or } 0\text{).}
\end{aligned}
$$

2. Let $\{v_1, \cdots, v_{n_\lambda}\}$ be a $K-$ basis of the left ideal $K[S_n]c_\lambda$ and extend it to a basis $\{v_1, \cdots, v_{n_\lambda}, \cdots, v_{n!}\}$ of $K[S_n]$. We have $v_i = x_i c_\lambda$ for some $x_i \in K[S_n]$, $1 \leq i \leq n_\lambda$. Hence we get that $v_i c_\lambda = x_i c_\lambda^2 = ax_i c_\lambda = av_i$ for all $1 \leq i \leq n_\lambda$. On the other hand, we have $v_j c_\lambda \in K[S_n]c_\lambda$ for all

$j > n_\lambda$. Thus the matrix of r_{c_λ} with respesct to this basis is of the following block form, namely,

$$\begin{pmatrix} A & B \\ 0 & 0 \end{pmatrix}, \quad \text{where } A = \text{diag}(a, \cdots, a),$$

consequently, we get that $\text{trace}(r_{c_\lambda}) = an_\lambda$. But then we get that $an_\lambda = n! \neq 0$ in K which implies that $a \neq 0$, as required.   ◇

5.5.6 Corollary: *We have $c_\lambda K[S_n]c_\lambda = Kc_\lambda$.*

For, let $x \in K[S_n]$ and consider $y = c_\lambda x c_\lambda$. By (5.5.1)((4) and (5)) above, we have $\sigma y \tau = \text{sgn}(\tau)y$ for all $\sigma \in R_\lambda$ and $\tau \in C_\lambda$ and hence we get $y = ac_\lambda$ for some $a \in K$. Thus we have $c_\lambda K[S_n]c_\lambda \subseteq Kc_\lambda$. On the other hand, we have $0 \neq c_\lambda^2$ is in $c_\lambda K[S_n]c_\lambda$ and so $c_\lambda K[S_n]c_\lambda = Kc_\lambda$, as required.   ◇

5.5.7 Theorem (Frobenius–Young): *The left ideal $V_\lambda = K[S_n]c_\lambda$ is minimal in $K[S_n]$ and hence affords an irreducible representation of S_n, called the Frobenius-Young module associated to λ.*

Proof: Since $c_\lambda \neq 0$, V_λ is a non–zero left ideal of $K[S_n]$.   Let $I \subseteq V_\lambda$ be a left ideal in $K[S_n]$. Now by (5.5.6) above, we have

$$c_\lambda I \subseteq c_\lambda(K[S_n]c_\lambda) = Kc_\lambda$$

which is 1-dimensional and hence we get that $c_\lambda I = (0)$ or else $c_\lambda I = Kc_\lambda$. Let, if possible, $c_\lambda I = (0)$. Then, we have $I^2 \subseteq (K[S_n]c_\lambda)I = K[S_n](c_\lambda I) = (0)$, i.e., I is a non–zero nilpotent ideal in the semi-simple ring $K[S_n]$ which is not possible. Hence $c_\lambda I = Kc_\lambda$. But then we have $c_\lambda \in Kc_\lambda = c_\lambda I \subseteq I$ and so $V_\lambda = K[S_n]c_\lambda \subseteq I \subseteq V_\lambda$, i.e., $I = V_\lambda$, as required.   ◇

5.5.8 Theorem: *For all distinct partitions λ and μ of n, we have*

$$a_\lambda K[S_n]b_\mu = (0) \quad \text{and} \quad c_\lambda K[S_n]c_\mu = (0).$$

Proof: Let $\lambda \neq \mu$. Since the dictionary order (5.2.2) $\succeq$ on $P(n)$ is a total order, we may assume that $\lambda \succ \mu$.

Step 1. We have $a_\lambda b_\mu = 0$.

For, by (5.5.2) above, there exists a pair (i, j) which occur in the same row of T_λ and in the same column of T_μ. Let $u = (i, j)$ so that $u \in R_\lambda \cap C_\mu$. By (5.5.1) above, we have

$$a_\lambda b_\mu = a_\lambda u u b_\mu = (a_\lambda u)(u b_\mu) = a_\lambda(-b_\mu) = -a_\lambda b_\mu = 0 \quad \text{as} \quad \text{Char } K \neq 2.$$

Step 2. We have $a_\lambda K[S_n] b_\mu = 0$.

For, we have $a_\lambda \theta b_\mu = a_\lambda \theta b_\mu \theta^{-1} \theta = a_\lambda b'_\mu \theta$ where $b'_\mu = \theta b_\mu \theta^{-1}$ which is associated to the Young tableaux $T'_\mu = \theta T_\mu$. But then by Step 1. we get that $a_\lambda b'_\mu = 0$ and so $a_\lambda \theta b_\mu = 0$.

Step 3. We have $c_\lambda x c_\mu = 0$ for all $x \in K[S_n]$.

For, given $x \in K[S_n]$, write $b_\lambda x a_\mu = \sum_{\theta \in S_n} a_\theta \theta$, $a_\theta \in K$. Using Step 2, we get that $c_\lambda x c_\mu = a_\lambda(b_\lambda x a_\mu)b_\mu = \sum_{\theta \in S_n} a_\theta(a_\lambda \theta b_\mu) = 0$. ◊

5.5.9 Corollary: *As S_n-modules, $V_\lambda \not\cong V_\mu$, $\forall \, \lambda \neq \mu$.*

For, we find that c_λ annihilates V_μ but not V_λ since $c_\lambda K[S_n] c_\mu = c_\lambda V_\mu = (0)$ but $c_\lambda^2 \neq 0$. Now the result follows since isomorphic modules must have their annihilator ideals equal. ◊

5.5.10 Corollary: *The family $\text{Irr}_K(S_n) = \{V_\lambda \mid \lambda \vdash n\}$ is a complete set of mutually inequivalent irreducible representations of S_n over K (by (5.5.7) and (5.5.9), above).*

5.5.11 Remark: $\text{Hom}_{S_n}(V_\lambda, V_\mu) = (0)$ for all $\lambda \neq \mu$.

This is immediate by Schur's lemma. Here is a direct verification. Let $f: V_\lambda \to V_\mu$ be an S_n-linear homomorphism. Let $f(c_\lambda) = x c_\mu$ for some $x \in K[S_n]$. Using (5.5.8) above, we have

$$f(c_\lambda K[S_n] c_\lambda) \subseteq c_\lambda K[S_n] f(c_\lambda) \subseteq c_\lambda K[S_n] x c_\mu \subseteq c_\lambda K[S_n] c_\mu = (0).$$

Since $a \neq 0$ and $0 \neq a c_\lambda = c_\lambda^2 \in c_\lambda K[S_n] c_\lambda$, we get that $a f(c_\lambda) = f(a c_\lambda) = f(c_\lambda^2) = 0$, resulting in $f(c_\lambda) = 0$, which implies that $f \equiv 0$.

5.5.12 Remark: In constructing the Frobenius–Young module V_λ, we have used a fixed Young tableaux T_λ of shape λ, i.e., $V_\lambda = V_\lambda(T_\lambda)$. If $T'_\lambda = \theta T_\lambda$ is another Young tableaux and $V'_\lambda = V_\lambda(T'_\lambda)$, then by (5.4.6) above, it follows that $V'_\lambda = \theta V_\lambda \theta^{-1}$, i.e., V_λ and V'_λ are equivalent representations of S_n.

5.5.13 Remark: The group of characters of S_n is given by $\widetilde{S_n} = \{(V_{(n)}, \iota), (V_{(1^n)}, \varepsilon)\}$. In fact, we have $V_{(n)} = K c_{(n)}$ is a one–dimensional 2–sided ideal on which S_n acts trivially. Hence the representation afforded is ι, the trivial character. On the other hand, we have $V_{(1^n)} = K c_{(1^n)}$ is also a one–dimensional 2–sided ideal on which S_n acts by the sgn character ε, etc.   ∎

5.6   III. Specht Modules for S_n

In this section, we shall give the Specht construction of the irreducible representations of S_n.

5.6.1 Let $K_n = K[X_1, \cdots, X_n]$ be the polynomial algebra over K in the n variables $X_1, \cdots, X_n$. We know that the group S_n acts (linearly) on K_n by permuting the variables, i.e., $\theta(X_i) = X_{\theta(i)}$, or $\theta(f(X_1, \cdots, X_n)) = f(X_{\theta(1)}, \cdots, X_{\theta(n)})$ for all $\theta \in S_n$ and $f \in K_n$.

For each positive integer m, let H_m denote the space of all homogeneous polynomials of degree m in K_n which is a finite dimensional vector subspace of K_n. Each H_m is an S_n–submodule of K_n.

5.6.2 Given a partition $\lambda \vdash n$, let λ' be the conjugate of λ. Let $\lambda = (\lambda_1, \cdots, \lambda_r)$ and $\lambda' = (\lambda'_1, \cdots, \lambda'_s)$. Let $m_\lambda = (1/2) \sum_{j=1}^{s} \lambda'_j(\lambda'_j - 1)$. Let T_λ be a fixed Young tableaux of shape λ. Let $a_{1j}, \cdots, a_{\lambda'_j j}$ be the entries of the jth column of T_λ. Define

$$\Delta_j = \Delta_j(a_{1j}, \cdots, a_{\lambda'_j j}) = \begin{vmatrix} 1 & \cdots & \cdots & 1 \\ X_{a_{1j}} & \cdots & \cdots & X_{a_{\lambda'_j j}} \\ X^2_{a_{1j}} & \cdots & \cdots & X^2_{a_{\lambda'_j j}} \\ \vdots & \vdots & \vdots & \vdots \\ X^{\lambda'_j - 1}_{a_{1j}} & \cdots & \cdots & X^{\lambda'_j - 1}_{a_{\lambda'_j j}} \end{vmatrix}$$

We note that this is a Vandermonde determinant and so we know that

$$\Delta_j(a_{1j}, \cdots, a_{\lambda'_j j}) = \prod_{1 \le p < q \le \lambda'_j} (X_{a_{qj}} - X_{a_{pj}})$$

which is a homogeneous polynomial of degree $m_\lambda = (1/2)\lambda'_j(\lambda'_j - 1)$.

5.6.3 Specht polynomials: With notation as in (5.6.2) above, we define $\Delta(T_\lambda) = \prod_{j=1}^{\lambda_1} \Delta_j$. This is a homogeneous polynomial of degree m_λ, called the *Specht polynomial* associated to the Young tableaux T_λ. (For $\sigma \in S_n$, we have $\sigma\Delta(T_\lambda) = \Delta(\sigma T_\lambda)$).

Examples: 1. Let $\lambda = (n)$. Then $T_{(n)}$ is a single row of n boxes filled with a permutation and $\Delta_1 = \Delta_2 = \cdots = \Delta_n = 1$ and hence $\Delta(T_{(n)}) = 1$ is the (monic) constant polynomial. We have $\theta(\Delta(T_{(n)})) = \Delta(T_{(n)})$ for all $\theta \in S_n$.
2. Let $\lambda = (1^n)$. Then $T_{(1^n)}$ is a single column of n boxes filled along a permutation, say the identity, for simplicity. Now we have

$$\Delta(T_{(1^n)}) = \Delta_1(1, \cdots, n)$$

$$= \begin{vmatrix} 1 & \cdots & \cdots & 1 \\ X_1 & \cdots & \cdots & X_n \\ X_1^2 & \cdots & \cdots & X_n^2 \\ \vdots & \vdots & \vdots & \vdots \\ X_1^{n-1} & \cdots & \cdots & X_n^{n-1} \end{vmatrix} = \prod_{1 \le i < j \le n} (X_j - X_i).$$

It is well–known that for all $\theta \in S_n$, we have

$$\theta(\Delta(T_{(1^n)})) = \Delta_1(\theta(1), \cdots, \theta(n)) = \mathrm{sgn}(\theta)\Delta(T_{(1^n)}).$$

5.6.4 Specht modules: Given $\lambda \vdash n$, the cyclic S_n–submodule of H_{m_λ} generated by the Specht polynomial $\Delta(T_\lambda)$ is independent of the tableaux T_λ but depends only on the shape λ. It is called the *Specht module* associated to the partition λ and is denoted by W_λ.

In fact, it is easy to see that the Specht module W_λ is spanned as a K–vector space by the set of all Specht polynomials of shape λ.

5.6.5 Theorem: *For $\lambda \vdash n$, the Specht module W_λ is simple.*

Proof: Since $K[S_n]$ is semi–simple, W_λ is finite dimensional and K is algebraically closed, the result is equivalent to showing that $\text{End}_{S_n}(W_\lambda) = K$.

Let $f \in \text{End}_{S_n}(W_\lambda)$. If $\lambda = (n)$, we know that $\Delta(T_{(n)}) = 1$ and so $W_{(n)} = K$ from which the result is obvious. Assume that $\lambda \neq (n)$. Let $\lambda = (\lambda_1, \cdots, \lambda_r)$. We have $\lambda_1 < n$ or equivalently, at least one (say, ℓ^{th}) column of T_λ is of length larger than 1. Let i, j be two indices belonging to such a column of T_λ. Let $\tau = (i, j)$ so that $\tau \in C_\lambda$. We have $\tau\Delta_\ell = -\Delta_\ell$ (since $X_i - X_j$ is a factor of Δ_ℓ) and $\tau\Delta_k = \Delta_k$ for all $k \neq \ell$ and hence $\tau\Delta(T_\lambda) = -\Delta(T_\lambda)$. Now we have

$$\tau f(\Delta(T_\lambda)) = f(\tau\Delta(T_\lambda)) = -f(\Delta(T_\lambda))$$

which means that $X_i - X_j$ is a factor of the polynomial $f(\Delta(T_\lambda))$. This is true for all entries i, j of the ℓth column of T_λ, i.e., each factor $X_i - X_j$ of Δ_ℓ is a factor of $f(\Delta(T_\lambda))$. But Δ_ℓ is square free and so it follows that Δ_ℓ divides $f(\Delta(T_\lambda))$. This is true for all columns of lengths larger than 1 of T_λ. Since $\Delta_k = 1$ for all columns of lengths 1, we get that $\Delta(T_\lambda)$ divides $f(\Delta(T_\lambda))$. Since $\Delta(T_\lambda)$ and $f(\Delta(T_\lambda))$ are of the same degree, we get that $f(\Delta(T_\lambda)) = a\Delta(T_\lambda)$ for some $a \in K$. Since W_λ is a cyclic $K[S_n]$–module generated by $\Delta(T_\lambda)$ and f is S_n–linear, we get that $f = a(\text{id})$, as required. $\Diamond$

5.6.6 Theorem: *For $\lambda \neq \mu$, the Specht modules W_λ and W_μ are non–isomorphic, i.e., $\text{Irr}_K(S_n) = \{W_\lambda \mid \lambda \vdash n\}$ is a complete set of mutually inequivalent irreducible representations of S_n over K.*

Proof: Since isomorphic modules have their annihilator ideals equal, it suffices to show that $\text{Ann}_{K[S_n]}W_\lambda \neq \text{Ann}_{K[S_n]}W_\mu$. We may assume that $\lambda \succ \mu$. We shall complete the proof in 4 steps.

1. If $(a_{1k}, \cdots, a_{\lambda'_k k})$ are the entries of the kth column of T_λ, we denote by C_k the subgroup of S_n permuting the entries of the kth column among themselves and all others fixed. Let $b_k = \sum_{\sigma \in C_k} \text{sgn}(\sigma)\sigma$. It is easy to see that $C_\lambda = C_1 \cdot C_2 \cdots C_{\lambda_1}$ and $b_\lambda = b_1 \cdot b_2 \cdots b_{\lambda_1}$.

We keep similar notation R_k, a_k, etc., for the corresponding entities of the row group and find that $R_\lambda = R_1 \cdot R_2 \cdots R_{\lambda'_1}$ and $a_\lambda =$

$a_1 \cdot a_2 \cdots a_{\lambda_1'}.$

2. We have $\Delta(T_\lambda) = b_\lambda \Phi(T_\lambda)$ for some monomial $\Phi(T_\lambda)$ in the X_i's.

By definition of the Specht polynomial, we have

$$
\Delta(T_\lambda) = \prod_{k=1}^{\lambda_1} \Delta_k(a_{1k}, \cdots, a_{\lambda_k' k})
$$

$$
= \prod_{k=1}^{\lambda_1} \begin{vmatrix} X^0_{a_{1k}} & \cdots & \cdots & X^0_{a_{\lambda_k' k}} \\ X_{a_{1k}} & \cdots & \cdots & X_{a_{\lambda_k' k}} \\ X^2_{a_{1k}} & \cdots & \cdots & X^2_{a_{\lambda_k' k}} \\ \vdots & \vdots & \vdots & \vdots \\ X^{\lambda_k'-1}_{a_{1k}} & \cdots & \cdots & X^{\lambda_k'-1}_{a_{\lambda_k' k}} \end{vmatrix}
$$

$$
= \prod_{k=1}^{\lambda_1} \Big(\sum_{\sigma \in C_k} \mathrm{sgn}(\sigma) \big(\prod_{\ell=0}^{\lambda_k'-1} X^\ell_{\sigma(a_{(\ell+1)k})} \big) \Big)
$$

$$
= \prod_{k=1}^{\lambda_1} \Big(\sum_{\sigma \in C_k} \mathrm{sgn}(\sigma) \sigma \big(\prod_{\ell=0}^{\lambda_k'-1} X^\ell_{a_{(\ell+1)k}} \big) \Big)
$$

$$
= \prod_{k=1}^{\lambda_1} \Big(\big(\sum_{\sigma \in C_k} \mathrm{sgn}(\sigma) \sigma \big) \big(\prod_{\ell=0}^{\lambda_k'-1} X^\ell_{a_{(\ell+1)k}} \big) \Big)
$$

$$
= \prod_{k=1}^{\lambda_1} \big(b_k \prod_{\ell=0}^{\lambda_k'-1} X^\ell_{a_{(\ell+1)k}} \big) = \big(\prod_{k=1}^{\lambda_1} b_k \big) \big(\prod_{k=1}^{\lambda_1} \prod_{\ell=0}^{\lambda_k'-1} X^\ell_{a_{(\ell+1)k}} \big)
$$

$$
= b_\lambda \big(\prod_{k=1}^{\lambda_1} \prod_{\ell=0}^{\lambda_k'-1} X^\ell_{a_{(\ell+1)k}} \big) = b_\lambda \prod_{k=1}^{\lambda_1} X^0_{a_{1k}} \cdot X^1_{a_{2k}} \cdots X^{\lambda_k'-1}_{a_{\lambda_k' k}}
$$

$$
= b_\lambda X^{\ell_1}_1 \cdot X^{\ell_2}_2 \cdots X^{\ell_n}_n = b_\lambda \Phi(T_\lambda), \quad \text{(say)}.
$$

We observe that the non–negative exponents ℓ_i's in the monomial $\Phi(T_\lambda)$ can easily be computed. For instance, $\ell_i = k - 1$ if and only if i occurs in the kth row of T_λ. Consequently, it follows that $\sigma\Phi(T_\lambda) = \Phi(T_\lambda)$ for all $\sigma \in R_\lambda$ and hence we get that $a_\lambda \Phi(T_\lambda) = \sum_{\sigma \in R_\lambda} \sigma\Phi(T_\lambda) = r_\lambda \Phi(T_\lambda)$ where r_λ is the order of the row group R_λ.

3. Now we have

$$
\begin{aligned}
a_\lambda \Delta(T_\lambda) &= a_\lambda b_\lambda \Phi(T_\lambda) = a_\lambda \Big(1 + \sum_{1 \neq \tau \in C_\lambda} \mathrm{sgn}(\tau)\tau \Big) \Phi(T_\lambda) \\
&= a_\lambda \Phi(T_\lambda) + a_\lambda \Big(\sum_{1 \neq \tau \in C_\lambda} \mathrm{sgn}(\tau)\tau \Big) \Phi(T_\lambda) \\
&= r_\lambda \Phi(T_\lambda) + \Big(\sum_{\sigma \in R_\lambda} \sum_{1 \neq \tau \in C_\lambda} \mathrm{sgn}(\tau)\sigma\tau \Big) \Phi(T_\lambda).
\end{aligned}
$$

Since $\sigma\tau\Phi(T_\lambda) \neq \Phi(T_\lambda)$ for all $\sigma \in R_\lambda$ and $1 \neq \tau \in C_\lambda$ and $r_\lambda \neq 0$ in K, we get that $a_\lambda \Delta(T_\lambda) \neq 0$, i.e., $a_\lambda \notin \mathrm{Ann}_{K[S_n]}(W_\lambda)$.

4. We shall now show that $a_\lambda \in \mathrm{Ann}_{K[S_n]}(W_\mu)$.

4(a). Since $\lambda \succ \mu$, there is a transposition $\tau = (i,j) \in R_\lambda \cap C_\mu$, by (5.5.2) above. We have $\tau \in R_k$ for some k. Since τ is disjoint from the entries of all other rows of T_λ, we get that τ commutes with every element of R_i for all $i \neq k$ and hence we have

$$
R_\lambda = R_1 \cdots R_{k-1} \cdot R_{k+1} \cdots R_{\lambda_1'} \cdot R_k.
$$

Consequently, we get that

$$
a_\lambda = a_1 \cdots a_{k-1} \cdot a_{k+1} \cdots a_{\lambda_1'} \cdot a_k = a_\lambda' \cdot a_k, \quad (\text{say}).
$$

4(b). We have $a_\lambda \Delta(T_\mu) = 0$.

Let A_k be the set of all even permutations in R_k. Then we have $R_k = A_k \cup \tau A_k$ which is a disjoint union. Let $\bar{a}_k = \sum_{\sigma \in A_k} \sigma$ so that $a_k = \bar{a}_k(1 + \tau) = (1 + \tau)\bar{a}_k$. Furthermore, we have $(1 + \tau)\Delta(T_\mu) = 0$ since $\tau \in C_\mu$ and so $\tau\Delta(T_\mu) = -\Delta(T_\mu)$. This gives that

$$
a_\lambda \Delta(T_\mu) = a_\lambda' \cdot a_k \Delta(T_\mu) = a_\lambda' \bar{a}_k \big((1 + \tau)\Delta(T_\mu) \big) = 0.
$$

4(c). We have $a_\lambda W_\mu = 0$.

$$
\begin{aligned}
a_\lambda W_\mu &= a_\lambda K[S_n]\Delta(T_\mu) = a_\lambda K[S_n](b_\mu \Phi(T_\mu)) \quad (\text{by } 2) \\
&= (a_\lambda K[S_n]b_\mu)\Phi(T_\mu) = 0 \cdot \Phi(T_\mu) = 0 \quad (\text{by } (5.5.8)). \qquad \diamond
\end{aligned}
$$

5.6.7 Remark: By (5.5.13), (5.6.3) (1) and (2) above, we find that the group of characters of S_n is given by

$$
\widetilde{S_n} = \{ (W_{(n)}, \iota), (W_{(1^n)}, \varepsilon) \} = \{ (V_{(n)}, \iota), (V_{(1^n)}, \varepsilon) \}. \qquad \blacksquare
$$

5.7 Standard Young Tableaux

In this section, we shall construct natural bases for the irreducible representations of S_n. The description being intrinsic, we can use the Frobenius–Young modules or the Specht modules. We shall use the Specht modules.

5.7.1 Standard Young tableaux: Given a partition $\lambda \vdash n$, a Young tableaux T_λ of shape λ is said to be *standard* if the entries in each row (resp. column) are in the increasing order from left to right (resp. top to bottom).

5.7.2 Examples of standard Young tableaux:
1. Let $\lambda = (7, 5, 2) \vdash 14$ and $T_\lambda = T_\lambda(\mathrm{id})$. Then T_λ is a standard tableaux where

T_λ :

1	4	7	9	11	13	14
2	5	8	10	12		
3	6					

2. For $\lambda = (3, 2) \vdash 5$, the following 5 tableaux are all the possible standard tableaux of shape $(3, 2)$.

1	3	5
2	4	

1	3	4
2	5	

1	2	5
3	4	

1	2	4
3	5	

1	2	3
4	5	

5.7.3 Standard Specht polynomials: The polynomial $\Delta(T_\lambda)$ associated to a standard tableaux T_λ is called a *standard Specht polynomial*.

Recall that the polynomial $\Delta(T_\lambda)$ associated to T_λ is a product of several determinants, namely, $\Delta(T_\lambda) = \prod_{k=1}^{\lambda_1} \Delta_k(a_{1k}, \cdots, a_{\lambda'_k k})$, where $a_{1k}, \cdots, a_{\lambda'_k k}$ are the entries of the kth column of T_λ and

$$\Delta_k(a_{1k}, \cdots, a_{\lambda'_k k}) = \begin{vmatrix} X^0_{a_{1k}} & \cdots & \cdots & X^0_{a_{\lambda'_k k}} \\ X_{a_{1k}} & \cdots & \cdots & X_{a_{\lambda'_k k}} \\ X^2_{a_{1k}} & \cdots & \cdots & X^2_{a_{\lambda'_k k}} \\ \vdots & \vdots & \vdots & \vdots \\ X^{\lambda'_k-1}_{a_{1k}} & \cdots & \cdots & X^{\lambda'_k-1}_{a_{\lambda'_k k}} \end{vmatrix}.$$

5.7.4 Diagonal terms of Specht polynomials: The product of the leading diagonals of the determinants (Δ_i's) defining a Specht polynomial $\Delta(T_\lambda)$ (which is a term in its expansion) is called the *diagonal term* or the *leading term* and is denoted by $D(T_\lambda)$.

For $\sigma \in S_n$ and $\lambda \vdash n$, let $T_\lambda(\sigma)$ be the tableaux filled along σ, i.e., filled with $\sigma(1), \cdots, \sigma(\lambda'_1)$ down the first column, $\sigma(\lambda'_1 + 1), \cdots, \sigma(\lambda'_1 + \lambda'_2)$ down the second column, etc. Then we have

$$\begin{aligned} D(T_\lambda(\sigma)) &= \left(X^0_{\sigma(1)} \cdots X^{\lambda'_1-1}_{\sigma(\lambda'_1)}\right) \cdots \left(X^0_{\sigma(n-\lambda'_s+1)} \cdots X^{\lambda'_s-1}_{\sigma(n)}\right) \\ &= \left(X_{\sigma(2)} \cdots X_{\sigma(n-\lambda'_s+2)}\right) \cdots \left(X^{\lambda'_1-1}_{\sigma(\lambda'_1)} \cdots X^{\lambda'_1-1}_{\sigma(\lambda_r\cdot\lambda'_1)}\right) \\ &= \left(X_{a_{21}} \cdots X_{a_{2\lambda_2}}\right)\left(\cdots\right)^2 \cdots \left(X_{a_{\lambda'_1 1}} \cdots X_{a_{\lambda'_1,\lambda_r}}\right)^{\lambda'_1-1} \\ &= P_1 P_2 \cdots P_{\lambda'_1-1}, \quad \text{(say)}, \end{aligned}$$

where a_{ij} is the entry of ith row j^{th} column of the tableaux $T_\lambda(\sigma)$ and P_i is   the ith power   of   the   monomial $(X_{a_{(i+1)1}} \cdots X_{a_{(i+1)\lambda(i+1)}})$, $1 \le i \le \lambda'_1 - 1$.

Basis Theorem for Specht Modules for S_n

5.7.5 Theorem: *The set of standard Specht polynomials of shape λ is a basis for the Specht module W_λ.*

Proof: It is but natural that the proof consists of the 2 main steps:

Step 1. *Linear independence of standard Specht polynomials.* (This is achieved by means of (5.7.6) to (5.7.8) below.)

Step 2. *Standard Specht polynomials span W_λ as a vector space.* (This is done in (5.7.9) through (5.7.11) below.)

The details, being given in full, appear to be a little lengthy (spread over 9 pages) but piecewise elementary. The essence of the proof is summarised in (5.7.13) below, to say that it is no more than verifying a *"formal setting"* as in (3.2.7) above.

Proof of Step 1: (5.7.6) to (5.7.8)

5.7.6 Lemma: *The diagonal terms of standard Specht polynomials of shape $\lambda \vdash n$ are linearly independent.*

Proof: Since the dioganal terms $D(T_\lambda)$'s are monomials of the same degree m_λ, the result is equivalent to showing that $D(T_\lambda) \neq D(T'_\lambda)$ for standard tableaux $T_\lambda \neq T'_\lambda$. This is trivial to see because we have

$$D(T_\lambda) = P_1 P_2 \cdots P_{\lambda'_1 - 1} = P'_1 P'_2 \cdots P'_{\lambda'_1 - 1} = D(T'_\lambda),$$

implying that $P_k = P'_k$ (for all k) for degree reasons. Hence if $T_\lambda = (a_{ij})$ and $T'_\lambda = (a'_{ij})$, we get that $(a_{k1}, \cdots, a_{k\lambda'_k})$ is a permutation of $(a'_{k1}, \cdots, a'_{k\lambda'_k})$ for all k. But both sets are in the ascending order by the standardness of T_λ and T'_λ which means that $a_{kj} = a'_{kj}$ for all j. This is true for all k and so we get that $T_\lambda = T'_\lambda$, as required. $\Diamond$

5.7.7 Lemma: 1. *For all $\sigma \in S_n$, we have $\sum_{j=1}^n j^2 \geq \sum_{j=1}^n j\sigma(j)$.*
2. *If $1 \leq a_1 < \cdots < a_t \leq n$ and S_t is the subgroup of S_n on these t symbols, then for all $\sigma \in S_t$, $\sigma \neq 1$, we have $\sum_{j=1}^t ja_j \geq \sum_{j=1}^t j\sigma(a_j)$.*

Proof: Since $\sum_{j=1}^n j^2 = \sum_{j=1}^n \sigma(j)^2$, we find that

$$2\sum_{j=1}^n \left(j^2 - j\sigma(j)\right) = \sum_{j=1}^n \left(j^2 + \sigma(j)^2 - 2j\sigma(j)\right) = \sum_{j=1}^n \left(j - \sigma(j)\right)^2 \geq 0,$$

proving the first part. We prove the second part by induction on t. If $t = 2$ or σ is a transposition, we have $\sigma = (a_i, a_j)$ with $i < j$ and hence

we find that $\sum_{k=1}^{t} ka_k - \sum_{k=1}^{t} k\sigma(a_k) = a_j - a_i > 0$, as required. Let
$t \geq 3$ and assume the induction hypothesis for $k < t$. If $\sigma(a_t) = a_t$,
then $\sigma \in S_{t-1}$ and hence the result follows by induction. We may
assume that $\sigma(a_t) \neq a_t$ and σ is not a transposition. Now consider τ
$= \rho\sigma$ where $\rho = (a_t, \sigma(a_t))$ so that we have $\tau(a_t) = a_t$, i.e., $\tau \in S_{t-1}$ and
$\tau \neq 1$. But then by induction, we get that $\sum_{k=1}^{t-1} k\tau(a_k) < \sum_{k=1}^{t-1} ka_k$.
Adding ta_t on either side, we get that $\sum_{k=1}^{t} k\tau(a_k) < \sum_{k=1}^{t} ka_k$. Since
$\rho(\tau(a_t)) = \rho(a_t) < a_t$ and the result is true for a transposition, it
follows that $\sum_{k=1}^{t} k\rho\tau(a_k) < \sum_{k=1}^{t} k\tau(a_k) < \sum_{k=1}^{t} ka_k$ and hence we
get that $\sum_{k=1}^{t} k\rho\tau(a_k) < \sum_{k=1}^{t} ka_k$. But $\rho\tau = \sigma$. $\Diamond$

5.7.8 Lemma: *Standard Specht polynomials of shape λ are linearly
independent.*

Proof: We shall prove that any non–trivial dependency relation
among the standard Specht polynomials $\Delta(T_\lambda)$ gives rise to a non–
trivial dependency relation among their diagonal terms $D(T_\lambda)$. This
leads to a contradiction by (5.7.6) above.

For a monomial $M = X_1^{\ell_1} \cdots X_n^{\ell_n}$ in $K[X_1, \cdots, X_n]$, let us call the
non–negative integer $\nu(M) = \sum_{j=1}^{n} j\ell_j$, the *numerical weight* of M. It
is obvious that $\nu(MN) = \nu(M) + \nu(N)$ for all monomials M and N.
(The numerical weight $\nu(M)$ can be interpreted as the degree of the
monomial M with respect to the variable X_j being assigned the new
degree j (instead of degree 1), $1 \leq j \leq n$.)

For integers $1 \leq a_1 < \cdots < a_t \leq n$, consider the determinant

$$\Delta(a_1, \cdots, a_t) \;=\; \begin{vmatrix} 1 & \cdots & \cdots & 1 \\ X_{a_1} & \cdots & \cdots & X_{a_t} \\ X_{a_1}^2 & \cdots & \cdots & X_{a_t}^2 \\ \vdots & \vdots & \vdots & \vdots \\ X_{a_1}^{t-1} & \cdots & \cdots & X_{a_t}^{t-1} \end{vmatrix}$$

$$= \sum_{\sigma \in S_t} \mathrm{sgn}(\sigma) X_{a_{\sigma(1)}}^0 \cdot X_{a_{\sigma(2)}} \cdots X_{a_{\sigma(t)}}^{t-1}.$$

Thus $\Delta(a_1, \cdots, a_t)$ is a sum of monomials of numerical weights $\sum_{j=1}^{t}(j-1)a_{\sigma(j)}$ of which the maximum is attained by the leading diagonal term, i.e., corresponding to $\sigma = 1$, by (5.7.7) above.

Now let T be a Young tableaux of shape λ and $\Delta(T)$ be the associated Specht polynomial, i.e., $\Delta(T) = \Delta_1 \cdots \Delta_s$, etc. On expansion of $\Delta(T)$, it is obvious that each monomial occuring in the sum is a product of monomials one each from that of the Δ_j's. By the multiplicativity of the numerical weight, it follows that the term M of the highest weight $\nu(M)$ in $\Delta(T)$ is the product of the terms of highest weights in each of the Δ_j's. But this is simply the diagonal term $D(T)$ of the Specht polynomial $\Delta(T)$.

Now we observe that any dependency relation among the Specht polynomials $\Delta(T_j)$'s gives rise to one among the terms of any fixed numerical weight and in particular, one among their diagonal terms $D(T_j)$'s, as required. $\Diamond$

Proof of Step 2: (5.7.9) to (5.7.11)

Since the set of all Specht polynomials of shape λ span W_λ, it suffices to show that any polynomial $\Delta(T_\lambda)$ can be written as a linear combination of suitable *standard* $\Delta(T'_\lambda)$'s. In fact, we shall prove that it is an *integral* linear combination of the standard $\Delta(T'_\lambda)$'s.

We proceed by induction on T_λ under the *lexicographic order* on the set of all Young tableaux of shape λ, namely, by comparing the sums of the entries in each column starting with the first. The least element under this order is $T_\lambda(\mathrm{id})$ corresponding to the filling by the identity permutation as per our convention, i.e., fill T_λ with the integers $1, \cdots, \lambda'_1$ down the first column, with the integers $\lambda'_1 + 1, \cdots, \lambda'_1 + \lambda'_2$ down the second column, etc. Since this least element is already standard, the result holds. Assume by induction that the result holds for all tableaux $T_\lambda(\theta)$, $\theta \in S_n$, smaller than the tableaux T_λ that we started with.

We may assume that T_λ is not standard, say the standardness fails between the ℓth and $(\ell+1)^{\text{th}}$ columns of T_λ whose entries are as shown below. The inequality is $>$ at the qth place with q least.

$$a_1 \quad < \quad b_1$$

$$\vdots$$

$$a_{q-1} \quad < \quad b_{q-1}$$

$$a_q \quad > \quad b_q$$

$$a_{q+1} \quad \cdot \quad b_{q+1}$$

$$\cdot$$

$$a_k \quad \cdot \quad b_k$$

$$\cdot$$

$$a_t$$

where $t = \lambda'_\ell \geq \lambda'_{\ell+1} = k$.

Let S_{t+1} be the subgroup of S_n on the $t+1$ integers $b_1, \cdots, b_q$, $a_q, \cdots, a_t$ (which are in the ascending order). Let S' be the subset of *shuffle permutations* of *level* q on these integers, namely,

$$S' = \left\{ \sigma \in S_{t+1} \Big| \sigma = \begin{pmatrix} b_1 < & \cdots < & b_q & < & a_q < & \cdots < & a_t \\ b_1^\sigma < & \cdots < & b_q^\sigma & ; & a_q^\sigma < & \cdots < & a_t^\sigma \end{pmatrix} \right\}.$$

This is a complete set of coset representatives of the subgroup $S_q \cdot S_{t-q+1}$ of S_{t+1} (and has $\binom{t+1}{q}$ elements) where S_q is the subgroup of S_n on the symbols $B_{\ell+1} = \{b_1, \cdots, b_q\}$ and S_{t-q+1} is the subgroup on $A_\ell = \{a_q, \cdots, a_t\}$.

5.7.9 Garnir element: With A_ℓ, $B_{\ell+1}$ and S' as above, the element in $K[S_n]$ defined by $G(A_\ell, B_{\ell+1}) = \sum_{\sigma \in S'} \operatorname{sgn}(\sigma)\sigma$ is called a *Garnir element* of the ℓth and $(\ell+1)^{\text{th}}$ columns of T_λ corresponding to the choice of the subsets A_ℓ and $B_{\ell+1}$ of the respective columns.

More generally, for a pair of consecutive columns, say k^{th} and $(k+1)^{\text{th}}$, of T_λ, let A be a bottom segment of length q_k in the kth column and B be a top segment of length q_{k+1} in the $(k+1)^{\text{th}}$ column such that $q_k + q_{k+1} > \lambda'_k$. Let $S(A, B)'$ be a complete set of coset representatives of the subgroup $S_{q_k}(A) \cdot S_{q_{k+1}}(B)$ of $S_{q_k+q_{k+1}}(A \cup B)$. The element in $K[S_n]$ defined by $G(A, B) = \sum_{\sigma \in S(A,B)'} \operatorname{sgn}(\sigma)\sigma$ is called a *Garnir element* of these columns associated to the segments A and B.

Now we have the following relation between the Vandermonde determinants $\Delta_\ell = \Delta(a_1, \cdots, a_t)$ and $\Delta_{\ell+1} = \Delta(b_1, \cdots, b_k)$ which are factors of $\Delta(T_\lambda)$.

5.7.10 Garnir relation: *We have*

$$
\begin{aligned}
\Delta_\ell \Delta_{\ell+1} &= \Delta(a_1, \cdots, a_t)\Delta(b_1, \cdots, b_k) \\
&= -\left(\sum_{1 \neq \sigma \in S'} \operatorname{sgn}(\sigma)\Delta^\sigma(a_1, \cdots, a_t)\Delta^\sigma(b_1, \cdots, b_k) \right) \\
&= -\left(\sum_{1 \neq \sigma \in S'} \operatorname{sgn}(\sigma)\Delta_\ell^\sigma \Delta_{\ell+1}^\sigma \right),
\end{aligned}
$$

where $\Delta_\ell^\sigma = \Delta^\sigma(a_1, \cdots, a_t) = \Delta(a_1, \cdots, a_{q-1}, a_q^\sigma, \cdots, a_t^\sigma)$ *and* $\Delta_{\ell+1}^\sigma = \Delta^\sigma(b_1, \cdots, b_k) = \Delta(b_1^\sigma, \cdots, b_q^\sigma, b_{q+1}, \cdots, b_k)$.

In other words, the Garnir element $G(A_\ell, B_{\ell+1})$ annihilates $\Delta_\ell \Delta_{\ell+1}$ because the stated Garnir relation is the same as saying

$$
G(A_\ell, B_{\ell+1})\Delta_\ell \Delta_{\ell+1} = \sum_{\sigma \in S'} \operatorname{sgn}(\sigma)\Delta_\ell^\sigma \Delta_{\ell+1}^\sigma = 0.
$$

This is a special case of an identity between the minors of a matrix $X = (X_{ij})$ of indeterminates for the particular choice $X_{ij} = X_j^{i-1}$, $1 \leq i, j \leq n$, in Lemma (5.7.11) below. We shall assume the validity of the Garnir relation and complete the proof of Step2 and then take up the proof of Lemma (5.7.11).

Proof of Step2 (contd): We have

$$
\Delta(T_\lambda) = \Delta_1 \cdots \Delta_{\ell-1}(\Delta_\ell \Delta_{\ell+1})\Delta_{\ell+2} \cdots \Delta_{\lambda_1}.
$$

Substituting for $\Delta_\ell \Delta_{\ell+1}$ (by the Garnir relation), we get that

$$
\Delta(T_\lambda) = - \sum_{1 \neq \sigma \in S'} \operatorname{sgn}(\sigma)\Delta_1 \cdots \Delta_\ell^\sigma \Delta_{\ell+1}^\sigma \cdots \Delta_{\lambda_1} = - \sum_{1 \neq \sigma \in S'} \Delta(T_\lambda^\sigma)
$$

where $T_\lambda^\sigma = \sigma(T_\lambda)$ for all $\sigma \in S'$. But now we see that for $1 \neq \sigma \in S'$, we have $a_i^\sigma = b_j$ for some i and j and hence we get that

$$
a_1 + \cdots + a_t > a_1 + \cdots + a_{q-1} + a_q^\sigma + \cdots + a_t^\sigma,
$$

i.e., the sum of the entries of the ℓth column of T^σ_λ is smaller than that of T_λ. But the first $\ell - 1$ columns of both the tableaux T^σ_λ and T_λ are identical. This implies that T^σ_λ is smaller than T_λ in the *lexicographic order* on the tableaux of shape λ. Hence by induction, each term $\Delta(T^\sigma)$ in the sum above is an *integral* linear combination of the standard Specht polynomials and so is their sum $\Delta(T_\lambda)$, as required. This completes the proof of Step2 and hence of (5.7.5) modulo the following lemma.   ◇

5.7.11 Lemma: *For integers $m, \ell_1, \cdots, \ell_m$ between 1 and n, let $\Delta(\ell_1, \cdots, \ell_m)$ be the minor of the matrix $X = (X_{ij})$ formed by the first m rows and the column indices $\ell_1, \cdots, \ell_m$. Then for integers a_i's and b_j's between 1 and t, as above, we have the following identity:*

$$\Delta(a_1, \cdots, a_t)\Delta(b_1, \cdots, b_k)$$
$$= -\Big(\textstyle\sum_{1 \neq \sigma \in S'} \operatorname{sgn}(\sigma)(\Delta^\sigma(a_1, \cdots, a_t)\Delta^\sigma(b_1, \cdots, b_k))\Big),$$
where $\Delta^\sigma(a_1, \cdots, a_t) = \Delta(a_1, \cdots, a_{q-1}, a_q^\sigma, \cdots, a_t^\sigma)$
and $\Delta^\sigma(b_1, \cdots, b_k) = \Delta(b_1^\sigma, \cdots, b_q^\sigma, b_{q+1}, \cdots, b_k)$.

Proof: The required identity is the same as showing that

$$(\star): \qquad \sum_{\sigma \in S'} \operatorname{sgn}(\sigma)\Delta^\sigma(a_1, \cdots, a_t)\Delta^\sigma(b_1, \cdots, b_k) = 0.$$

We apply Laplace expansion of a determinant with respect to a specified set of rows or columns as the case may be.

Expanding with respect to the first $q - 1$ columns of the following determinant, we get that

$$\Delta^\sigma(a_1, \cdots, a_t) = \begin{vmatrix} X_{1a_1} & \cdots & X_{1a_{q-1}} & X_{1a_q^\sigma} & \cdots & X_{1a_t^\sigma} \\ \vdots & & & & & \vdots \\ X_{q-1a_1} & \cdots & X_{q-1a_{q-1}} & X_{q-1a_q^\sigma} & \cdots & X_{q-1a_t^\sigma} \\ X_{qa_1} & \cdots & X_{qa_{q-1}} & X_{qa_q^\sigma} & \cdots & X_{qa_t^\sigma} \\ \vdots & & & & & \vdots \\ X_{ta_1} & \cdots & X_{ta_{q-1}} & X_{ta_q^\sigma} & \cdots & X_{ta_t^\sigma} \end{vmatrix}$$

$$= \sum_{\tau \in S!} \operatorname{sgn}(\tau)\Phi_\tau \Psi^\sigma_\tau, \quad \text{(say)},$$

where S_1' is the set of shuffle permutations of level $q - 1$ on the set $1, \cdots, t$ and for each $\tau \in S_1'$, we have

$$\Phi_\tau = \Phi_\tau(a_1, \cdots, a_{q-1}) = \begin{vmatrix} X_{\tau(1)a_1} & \cdots & X_{\tau(1)a_{q-1}} \\ \vdots & & \vdots \\ X_{\tau(q-1)a_1} & \cdots & X_{\tau(q-1)a_{q-1}} \end{vmatrix}$$

and

$$\Psi_\tau^\sigma = \Psi_\tau^\sigma(a_q^\sigma, \cdots, a_t^\sigma) = \begin{vmatrix} X_{\tau(q)a_q^\sigma} & \cdots & X_{\tau(q)a_t^\sigma} \\ \vdots & & \vdots \\ X_{\tau(t)a_q^\sigma} & \cdots & X_{\tau(t)a_t^\sigma} \end{vmatrix}.$$

Likewise, if S_2' is the set of shuffle permutations of level q on the set $1, \cdots, k$, then on Laplace expansion with respect to the first q columns, we get that

$$\Delta^\sigma(b_1, \cdots, b_k) = \begin{vmatrix} X_{1b_1^\sigma} & \cdots & X_{1b_q^\sigma} & X_{1b_{q+1}} & \cdots & X_{1b_k} \\ \vdots & & & & & \vdots \\ X_{qb_1^\sigma} & \cdots & X_{qb_q^\sigma} & X_{qb_{q+1}} & \cdots & X_{qb_k} \\ X_{q+1b_1^\sigma} & \cdots & X_{q+1b_q^\sigma} & X_{q+1b_{q+1}} & \cdots & X_{q+1b_k} \\ \vdots & & & & & \vdots \\ X_{kb_1^\sigma} & \cdots & X_{kb_q^\sigma} & X_{kb_{q+1}} & \cdots & X_{kb_k} \end{vmatrix}$$

$$= \sum_{\theta \in S_2'} \operatorname{sgn}(\theta) \Psi_\theta^\sigma \Phi_\theta, \quad \text{(say)},$$

where for each $\theta \in S_2'$, we have

$$\Psi_\theta^\sigma = \Psi_\theta^\sigma(b_1^\sigma, \cdots, b_q^\sigma) = \begin{vmatrix} X_{\theta(1)b_1^\sigma} & \cdots & X_{\theta(1)b_q^\sigma} \\ \vdots & & \vdots \\ X_{\theta(q)b_1^\sigma} & \cdots & X_{\theta(q)b_q^\sigma} \end{vmatrix}$$

and

$$\Phi_\theta = \Phi_\theta(b_{q+1}, \cdots, b_k) = \begin{vmatrix} X_{\theta(q+1)b_{q+1}} & \cdots & X_{\theta(q+1)b_k} \\ \vdots & & \vdots \\ X_{\theta(k)b_{q+1}} & \cdots & X_{\theta(k)b_k} \end{vmatrix}.$$

Now substituting in the left hand side of $(\star)$, we get that

$$
\begin{aligned}
\text{LHS of } (\star) &= \sum_{\sigma \in S'} \sum_{\tau \in S_1'} \sum_{\theta \in S_2'} \operatorname{sgn}(\sigma)\operatorname{sgn}(\tau)\operatorname{sgn}(\theta)\Phi_\tau \Psi_\tau^\sigma \Psi_\theta^\sigma \Phi_\theta \\
&= \sum_{\tau \in S_1'} \sum_{\theta \in S_2'} \operatorname{sgn}(\tau)\operatorname{sgn}(\theta)\Phi_\tau \Phi_\theta \Big(\sum_{\sigma \in S'} \operatorname{sgn}(\sigma)\Psi_\theta^\sigma \Psi_\tau^\sigma \Big) \\
&= \sum_{\tau \in S_1'} \sum_{\theta \in S_2'} \operatorname{sgn}(\tau)\operatorname{sgn}(\theta)\Phi_\tau \Phi_\theta \Psi_{\theta\tau}, \quad \text{where} \\
\Psi_{\theta\tau} &= \sum_{\sigma \in S'} \operatorname{sgn}(\sigma)\Psi_\theta^\sigma \Psi_\tau^\sigma.
\end{aligned}
$$

But we see that each term $\Psi_{\theta\tau}$ above is nothing but the Laplace expansion of the following $(t+1) \times (t+1)$ determinant with respect to the first q rows, namely,

$$
\Psi_{\theta\tau} = \begin{vmatrix}
X_{\theta(1)b_1} & \cdots & X_{\theta(1)b_q} & X_{\theta(1)a_q} & \cdots & X_{\theta(1)a_t} \\
\vdots & & & & & \vdots \\
X_{\theta(q)b_1} & \cdots & X_{\theta(q)b_q} & X_{\theta(q)a_q} & \cdots & X_{\theta(q)a_t} \\
X_{\tau(q)b_1} & \cdots & X_{\tau(q)b_q} & X_{\tau(q)a_q} & \cdots & X_{\tau(q)a_t} \\
\vdots & & & & & \vdots \\
X_{\tau(t)b_1} & \cdots & X_{\tau(t)b_q} & X_{\tau(t)a_q} & \cdots & X_{\tau(t)a_t}
\end{vmatrix}.
$$

Finally, we find that this determinant is zero because row indices are the $t+1$ integers $\theta(1), \cdots, \theta(q), \tau(q), \cdots, \tau(t)$ between 1 and t and so at least two of them must be equal. $\Diamond$

5.7.12 Corollary: *Representations of S_n are defined over $\mathbb{Z}$, in the sense that given a representation (V, ρ) of S_n, there exists a basis of V with respect to which the matrices of $\rho(\sigma)$, $\sigma \in S_n$, have integer entries. In particular, the character of any representation of S_n is integer valued. (See also (5.3.4) above.)*

It suffices to show that any irreducible representation of S_n is defined over $\mathbb{Z}$. To see this, we use the standard Specht polynomial basis for the irreducible representation $(W_\lambda, \rho_\lambda)$ for $\lambda \vdash n$. From the proof of Step 2 of (5.7.5) above, we have for each $\sigma \in S_n$ and a standard Specht polynomial $\Delta(T_\lambda)$, that $\rho_\lambda(\sigma)\big(\Delta(T_\lambda)\big) = \sigma\Delta(T_\lambda) =$

$\Delta(\sigma T_\lambda)$ is either standard or an integral linear combination of the standard polynomials and hence the matrix of $\rho_\lambda(\sigma)$ with respect to this basis is an integral matrix, as required.

5.7.13 Remark: In proving (5.7.5) above, what we have essentially done is that we simply verified the hypothesis H_1, H_2 and H_3 of (3.2.7) above, for the cyclic S_n–module W_λ generated by the Specht polynomial $\Delta(T_\lambda) = \Delta(T_\lambda(1))$, i.e., we have done the following.

H$_1$: Identifying the group S_n with the set of all Young tableaux $T_\lambda(\theta)$ of shape λ, filled along $\theta \in S_n$, we have a total order '$\leq$' on S_n corresponding to the *lexicographic order* on these tableaux (used in Step 2). We have $1 \leq \theta$ for all θ in S_n.

H$_2$: If $X = \{\theta \in S_n \mid T_\lambda(\theta)$ is standard$\}$, then we have $1 \in X$ and verified that the subset of W_λ, namely, $X \cdot \Delta(T_\lambda(1)) = \{\Delta(T_\lambda(\theta)) \mid \theta \in X\}$ is linearly independent.

H$_3$: For $\theta \notin X$, if $g_\theta = \theta^{-1} G(A_\ell, B_{\ell+1})\theta$ where $G(A_\ell, B_{\ell+1})$ is the associated Garnir element as in (5.7.9) above, then we have $1 \in \operatorname{supp}(g_\theta)$ with the Garnir relation $g_\theta \Delta(T_\lambda(1)) = \theta^{-1} G(A_\ell, B_{\ell+1})\Delta(T_\lambda(\theta)) = 0$. Finally, we have $\operatorname{supp}(g_\theta) = \theta^{-1} S' \theta$ where S' is as in (5.7.9) above. Consequently, for all $\sigma' \in \operatorname{supp}(g_\theta)$, we have $\sigma' = \theta^{-1}\sigma\theta$ for some $\sigma \in S'$ and hence $\theta\sigma' = \sigma\theta \leq \theta$ in the total order on S_n as in H_1, above since the corresponding tableaux $T_\lambda(\theta)^\sigma = T_\lambda(\sigma\theta)$ is smaller than $T_\lambda(\theta)$ for all $\sigma \in S'$.

5.7.14 Remark: By similar cosiderations as above, it can be seen that the Frobenius–Young module V_λ has a basis consisting of $Xc_\lambda = \{\theta c_\lambda \mid T_\lambda(\theta)$ is standard$\}$. ■

5.8 Hook–Length Formula

In this section, we shall determine the dimensions of the (ordinary) irreducible representattions of S_n. From the basis Theorem (5.7.5) for the Specht modules W_λ, $\lambda \vdash n$, we have $d_\lambda := \dim_K W_\lambda$ is the number of standard Young tableaux of shape λ. This purely combinatorial problem of counting the number of standard tableaux is quite intriguing. The ultimate formula is very elegant and quite simple to apply in any particular case. However, a proof of the formula itself is not so simple. It involves an intricate machinery containing several formal polynomial and determinantal identities and so forth. These identities are interesting in their own right and not difficult to establish but technically involved, occupying considerable space. As a first encounter, it is perhaps better to assume the formula. However, we shall present an outline assuming two crucial facts. The interested reader should fill up the missing details, consulting the indicated sources. Now we state the formula and show how simple it is for applications.

5.8.1 Hook–arm: Let T_λ be a Young diagram of shape $\lambda \vdash n$. By the *hook-arm* of a box or a position of T_λ, we mean the union of the parts of the row and column of boxes lying below or to the right of and inclusive of that box.

The hook–arms of the $(2,2)^{\text{nd}}$ and $(1,3)^{\text{rd}}$ positions of T_λ for $\lambda = (8,5,4,3^2,1) \vdash 24$ are shown below.

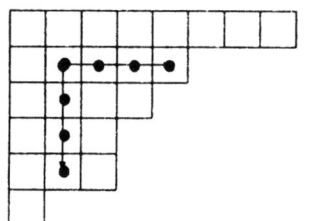 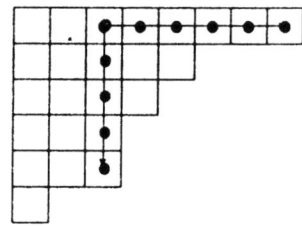

5.8.2 Hook–length of a position: The *hook-length* of the $(j,k)^{\text{th}}$ position of a Young diagram is defined as the length of the hook–arm of that position (i.e., the number of boxes or •'s the hook–arm passes through) and is denoted by $h_{jk}(\lambda)$.

If $\lambda = (\lambda_1, \cdots, \lambda_r)$ and its conjugate $\lambda' = (\lambda'_1, \cdots, \lambda'_s)$, then we have $h_{jk}(\lambda) = \lambda_j + \lambda'_k - (j+k) + 1$.

In the examples shown above, we have $h_{22}(\lambda) = 7$ and $h_{13}(\lambda) = 10$ where $\lambda = (8,5,4,3^2,1)$ and its conjugate $\lambda' = (6,5^2,3,2,1^3)$.

5.8.3 Hook–length table: The *hook-length table* or the *hook-legnth graph* of a Young diagram of shape λ is defined to be the same Young diagram filled with the hook–lengths of the respective positions and is denoted by $H_\lambda = \big((h_{jk}(\lambda))\big)$, $1 \le j \le \lambda'_1$ and $1 \le k \le \lambda_1$.

The following is the hook–length table of shape $\lambda = (7, 5, 2) \vdash 14$.

$$H_\lambda :$$

9	8	6	5	4	2	1
6	5	3	2	1		
2	1					

5.8.4 Hook–length of a diagram: The *hook-length of a diagram* of shape λ, or simply the *hook–length* of λ, is defined to be the product of the hook–lengths of all of its positions and is denoted by h_λ, i.e.,

$$h_\lambda = \prod_{\substack{1 \le j \le \lambda'_1 \\ 1 \le k \le \lambda_1}} h_{jk}(\lambda).$$

For instance, we have

$$h_{(7,5,2)} = 9 \cdot 8 \cdot 6 \cdot 5 \cdot 4 \cdot 2 \cdot 1 \cdot 6 \cdot 5 \cdot 3 \cdot 2 \cdot 1 \cdot 2 \cdot 1 = 6,220,800.$$

5.8.5 Theorem (Hook–length formula): *For each partition $\lambda \vdash n$, the number d_λ of standard Young tableaux of shape λ is given by*

$$d_\lambda = \frac{n!}{h_\lambda},$$

called the hook-length formula, giving the dimension of the irreducible representation W_λ of S_n corresponding to the conjugacy class λ.

5.8.6 Examples: 1. For $\lambda = (3,2) \vdash 5$, we have seen in (5.7.2) above, that there are exactly 5 standard tableaux of shape (3,2). The hook–length table of shape (3,2) is the following.

$$H_{(3,2)} =$$

4	3	1
2	1	

Thus we have $h_{(3,2)} = 4 \cdot 3 \cdot 2$ and hence $d_{(3,2)} = 5!/4! = 5$, as we have already explicitly written out the standard tableaux in (5.7.2)(2) above.

2. The dimensions of all the inequivalent irreducible representations of S_5 are   1, 4, 5, 6, 5, 4 and 1   respectively arranged according to the descending order (5.2.2) of the partitions of 5.

3. For $\lambda = (7,5,2)$, we get from (5.8.3) above, that $h_{(7,5,2)} = 9 \cdot 8 \cdot 6 \cdot 5 \cdot 4 \cdot 2 \cdot 6 \cdot 5 \cdot 3 \cdot 2 \cdot 2 = 6{,}220{,}800$ and hence we have $d_{(7,5,2)} = 14!/6{,}220{,}800 = 7 \cdot 11 \cdot 13 \cdot 14 = 14{,}014$.

5.8.7   The rest of this section is devoted to an outline of the proof of the hook–length formula.

There is no quick proof available for this innocent looking formula. There is no direct way of seeing that the fraction $n!/h_\lambda$ *is an integer, let alone that it is the number of standard tableaux.*

We need some terminology and notation before we can outline a proof.

5.8.8 Semi–standard tableax: A Young tableaux of shape $\lambda \vdash n$, filled with the integers between 1 and n but *allowing repetitions*, is said to be *semi-standard* if the entries down the columns are strictly increasing but those in the rows are weakly increasing (from left).

5.8.9 Weight of a tableaux: Given a semi–standard tableaux T_λ, let ν_j be the number of times j occurs in T_λ so that we have $\nu(T_\lambda) = (\nu_1, \cdots, \nu_n) \in \mathbb{Z}^{+n}$ with $\sum_{j=1}^{n} \nu_j = n$. This $\nu(T_\lambda)$ is called the *weight* of the tableaux T_λ.

It is obvious that a *semi–standard tableaux is standard* if and only if its weight is $(1^n) = (1, \cdots, 1) \in \mathbb{N}^n$.

5.8.10 Monomial associated to a tableaux: Given a semi-standard tableaux T_λ with its weight $\nu(T_\lambda)$, the monomial $X^{\nu(T_\lambda)} = X_1^{\nu_1} \cdots X_n^{\nu_n}$ of   degree n in   the   polynomial   ring $K[X_1, \cdots, X_n]$ is called the *monomial associated* to the tableaux T_λ.

We note that a semi–standard tableaux is standard if and only if its monomial is $X_1 \cdots X_n$, i.e., the *square-free monomial* of maximum degree n.

5.8.11   Given a $\lambda \vdash n$, we define the *polynomial associated* to the set of semi–standard tableaux of shape λ to be $S'_\lambda = \sum_{T_\lambda} X^{\nu(T_\lambda)}$ which is homogeneous of degree n.

Obviously the coefficient of the monomial $X_1 \cdots X_n$ in the polynomial S'_λ is precisely d_λ, the number of standard tableaux of shape λ.

5.8.12   Symmetric and Skew–symmetric polynomials: A polynomial $f(X_1, \cdots, X_n)$ is said to be *symmetric* if

$$f(X_{\theta(1)}, \cdots, X_{\theta(n)}) = f(X_1, \cdots, X_n), \ \forall \ \theta \in S_n.$$

Likewise, $f(X_1, \cdots, X_n)$ is said to be *skew-symmetric* if

$$f(X_{\theta(1)}, \cdots, X_{\theta(n)}) = \text{sgn}(\theta) f(X_1, \cdots, X_n), \ \forall \ \theta \in S_n.$$

We recall the well–known fact that the set of all symmetric polynomials is a subalgebra of $K[X_1, \cdots, X_n]$ generated by the *elementary symmetric functions* $\{E_j \mid 1 \le j \le n\}$ where E_j is the sum of all *square-free monomials* of degree j. In fact, the E_j's are *algebraically independent*, i.e., like independent variables, and any symmetric polynomial is a polynomial in the E_j's.

A not so well–known fact is that the vector space of all skew-symmetric homogeneous polynomials of degree N has a basis consisting of the Vandermonde type determinants Δ_Λ defined below.

5.8.13   Let $\lambda \vdash n$ and write $\lambda = (\lambda_1, \cdots, \lambda_n)$ allowing a string of zeros as parts of λ. Let $\Lambda = (\Lambda_1, \cdots, \Lambda_n)$ where $\Lambda_j = \lambda_j + n - j$ so that Λ is a strictly decreasing partition (s.d.p) of $N = n(n+1)/2$ in the sense of (5.2.3) above. It is rather obvious that every strictly decreasing partition Λ of N having atmost n parts is of the form $\Lambda = \lambda + \delta$ (addition of parts coordinatewise) for a unique $\lambda \vdash n$ where $\delta = (n-1, n-2, \cdots, 1, 0) \vdash n(n-1)/2$.

5.8.14   Let Δ_Λ be the Vandermonde type determinant defined by

$$\Delta_\Lambda = \begin{vmatrix} X_1^{\Lambda_n} & \cdots & X_n^{\Lambda_n} \\ \vdots & \vdots & \vdots \\ X_1^{\Lambda_1} & \cdots & X_n^{\Lambda_1} \end{vmatrix}.$$

It is clear that Δ_Λ is skew–symmetric and therefore $X_j - X_k$ divides Δ_Λ for all $j \neq k$ and hence the fraction $S_\lambda = \Delta_\Lambda / \Delta_\delta$ is a homogeneous symmetric polynomial of degree n with integer coefficients where

$$\Delta_\delta = \prod_{1 \leq j < k \leq n} (X_k - X_j) = \begin{vmatrix} X_1^0 & \cdots & X_n^0 \\ \vdots & \vdots & \vdots \\ X_1^{n-1} & \cdots & X_n^{n-1} \end{vmatrix}$$

is the Vandermonde determinant.

5.8.15   Schur functions: Given a $\lambda \vdash n$, the homogeneous symmetric polynomial S_λ of degree n with integer coefficients, defined above, is called the *Schur function* or *S-function* associated to λ.

There are a host of K–bases for the space of all symmetric polynomials. One such important basis is the set of Schur functions associated to partitions.

The two crucial facts that clinch the proof of the hook–length formula are the following:

Fact 1. *The coefficient of the monomial $X_1 \cdots X_n$ in the Schur function S_λ is equal to the fraction $n!/h_\lambda$ and hence it is an integer.*

Fact 2. *The Schur function S_λ associated to λ is the same as the polynomial S'_λ associated to the set of all semi–standard tableaux of shape λ, defined in (5.8.11) above.*

Consequently, we get the hook–length formula that

$$d_\lambda = \text{coefficient of } X_1 \cdots X_n \text{ in } S'_\lambda$$

$$= \text{coefficient of } X_1 \cdots X_n \text{ in } S_\lambda = \frac{n!}{h_\lambda}, \quad \text{as required.} \qquad \Diamond$$

For a proof the Facts 1 and 2 above, see [35] pp. 23–43, [32] pp. 94–111, [24] p.77 and/or [25] pp. 56 ar. 1 304.   ■

5.9   IV. Irreducible Representations of S_n– An Abstract Method

In this section, we assume that Char $K = 0$. We shall realise the irreducible representations $\{V_\lambda \mid \lambda \vdash n\}$, given by the Frobenius–Young method (5.5.7), by an abstract method. This method is more suitable to obtain the irreducible representations of the alternating subgroups A_n (Chapter 6 below).

For $\lambda \vdash n$, let λ' be its conjugate. Fix a Young tableaux T_λ of shape λ. Let R_λ and C_λ be its row and column groups (5.4.4). It is more convenient to take $T_{\lambda'} = {}^t T_\lambda$, the transpose of T_λ so that $R_{\lambda'} = C_\lambda$ and so on.

Let ιR_λ be the trivial character (5.5.13) of R_λ. Likewise, let ϵC_λ be the "sgn"character of C_λ. Let $\mathrm{Ind}\,(\iota R_\lambda)\!\uparrow_{R_\lambda}^{S_n}$ and $\mathrm{Ind}\,(\epsilon C_\lambda)\!\uparrow_{C_\lambda}^{S_n}$ be the induced representations. Now we have the following.

5.9.1 Theorem: *The representations* $\mathrm{Ind}\,(\iota R_\lambda)\!\uparrow_{R_\lambda}^{S_n}$ *and* $\mathrm{Ind}\,(\epsilon C_\lambda)\!\uparrow_{C_\lambda}^{S_n}$ *of* S_n *have a unique irreducible subrepresentation of multiplicity 1 in common and it is denoted by* $[\lambda] = \mathrm{Ind}\,(\iota R_\lambda)\!\uparrow_{R_\lambda}^{S_n} \;\bigcap\; \mathrm{Ind}\,(\epsilon C_\lambda)\!\uparrow_{C_\lambda}^{S_n}$.

Proof: Since the base field is of characteristic 0, the result is equivalent to showing that $\left\langle \chi_{\mathrm{Ind}(\iota R_\lambda)\uparrow_{R_\lambda}^{S_n}} \,,\, \chi_{\mathrm{Ind}(\epsilon C_\lambda)\uparrow_{C_\lambda}^{S_n}} \right\rangle_{S_n} = 1$. Let us then find the inner product using the Frobenius reciprocity for characters (4.2.2) and the Subgroup Theorem (4.5.3).

Let D_λ be a complete set of double coset representatives of $R_\lambda \backslash S_n / C_\lambda$. Assume that the identity element $1 \in D_\lambda$. For $\pi \in D_\lambda$, let us write $C_{\lambda\pi} = R_\lambda \cap C_\lambda{}^\pi = R_\lambda \cap (\pi C_\lambda \pi^{-1})$. Note that we have
(i) $C_{\lambda 1} = (1)$ which gives that   (ii) $(\iota R_\lambda)\!\downarrow_1^{R_\lambda} = (\epsilon C_\lambda)\!\downarrow_1^{C_\lambda}$,
(iii) for $\pi \in D_\lambda, \pi \neq 1 \Rightarrow \pi \neq xy$, $\forall\, x \in R_\lambda$ and $y \in C_\lambda$ and hence
(iv) for $\pi \neq 1$, $C_{\lambda\pi}$ contains a transposition (by (5.5.3) above) and
(v) $(\iota R_\lambda)\!\downarrow_{C_{\lambda\pi}}^{R_\lambda}$ is *not equivalent* to $(\epsilon C_\lambda)\!\downarrow_{C_{\lambda\pi}}^{C_\lambda}$, i.e.,

$$\left\langle \chi_{(\iota R_\lambda)\downarrow C_{\lambda\pi}} \,,\, \chi_{(\epsilon C_\lambda)\downarrow C_{\lambda\pi}} \right\rangle_{C_{\lambda\pi}} = 0, \;\forall\, \pi \neq 1.$$

Now we have

$$\left\langle \chi_{\mathrm{Ind}(\iota R_\lambda)\uparrow_{R_\lambda}^{S_n}}, \chi_{\mathrm{Ind}(\epsilon C_\lambda)\uparrow_{C_\lambda}^{S_n}} \right\rangle_{S_n} = \left\langle \chi_{\iota R_\lambda}, \left(\chi_{\mathrm{Ind}(\epsilon C_\lambda)\uparrow_{C_\lambda}^{S_n}} \right)\downarrow_{R_\lambda}^{S_n} \right\rangle_{R_\lambda}$$

$$= \sum_{\pi \in D_\lambda} \left\langle \chi_{\iota R_\lambda}, \chi_{((\epsilon C_\lambda)^\pi \iota_{C_{\lambda\pi}})\uparrow_{C_{\lambda\pi}}^{R_\lambda}} \right\rangle_{R_\lambda}$$

$$= \sum_{\pi \in D_\lambda} \left\langle \chi_{(\iota R_\lambda)\downarrow_{C_{\lambda\pi}}}, \chi_{(\epsilon C_\lambda)^\pi \iota_{C_{\lambda\pi}}} \right\rangle_{C_{\lambda\pi}}$$

$$= \sum_{\pi \in D_\lambda} \left\langle \chi_{(\iota R_\lambda)\downarrow_{C_{\lambda\pi}}}, \chi_{(\epsilon C_\lambda)\downarrow_{C_{\lambda\pi}}} \right\rangle_{C_{\lambda\pi}}$$

$$= 1 + \sum_{\pi \neq 1} \left\langle \chi_{(\iota R_\lambda)\downarrow_{C_{\lambda\pi}}}, \chi_{(\epsilon C_\lambda)\downarrow_{C_{\lambda\pi}}} \right\rangle_{C_{\lambda\pi}}$$

$$= 1 + 0, \quad \text{as required.} \qquad \diamond$$

5.9.2 Examples: It is easy to see that we have
(i) $[(n)] = \iota S_n$ and **(ii)** $[(1^n)] = \epsilon S_n$.

In fact, these two are special cases of the following.

5.9.3 Corollary: *For all* $\lambda \vdash n$, *we have*
(i) $[\lambda'] = [\lambda] \otimes (\epsilon S_n)$ *and consequently,* **(ii)** $[\lambda]\downarrow_{A_n}^{S_n} \equiv [\lambda']\downarrow_{A_n}^{S_n}$.

Proof: Since $(V \otimes_K W)\downarrow_H^G = (V\downarrow_H^G) \otimes_K (W\downarrow_H^G))$, we have only to see
the first which follows from the obvious facts that
(i) $\mathrm{Ind}\,(\iota)\uparrow_1^{S_n} = K[S_n]$, **(ii)** $\mathrm{Ind}\,(\epsilon)\uparrow_{S_n}^{S_n} = \epsilon S_n$,
(iii) $\mathrm{Ind}\,(\iota R_\lambda)\uparrow_{R_\lambda}^{S_n} \otimes(\epsilon S_n) = \mathrm{Ind}\,(\epsilon R_\lambda)\uparrow_{R_\lambda}^{S_n}$ and
(iv) $\mathrm{Ind}\,(\iota C_\lambda)\uparrow_{C_\lambda}^{S_n} = \mathrm{Ind}\,(\epsilon C_\lambda)\uparrow_{C_\lambda}^{S_n} \otimes(\epsilon S_n)$. $\qquad \diamond$

We shall now prove that $[\lambda] \not\cong [\mu]$ for all $\lambda \neq \mu$. To do this,
recall the notation (from §5.5 above), namely, $a_\lambda = \sum_{\sigma \in R_\lambda} \sigma$, $\quad b_\lambda = \sum_{\tau \in C_\lambda} \mathrm{sgn}(\tau)\tau$ and $c_\lambda = a_\lambda b_\lambda$ and the facts that $c_\lambda \neq 0$ (5.5.1) and $V_\lambda = K[S_n]c_\lambda$ is a minimal left ideal of $K[S_n]$ (5.5.7), etc.

5.9.4 Theorem: *For all* $\lambda \vdash n$, *we have* $[\lambda] = K[S_n]c_\lambda$ *and hence
the family* $\mathrm{Irr}_K(S_n) = \{ [\lambda] \mid \lambda \vdash n\}$ *is a complete set of irreducible
representations of* S_n.

Proof: By Ex.(4.8.8) above, we have $\mathrm{Ind}\,(\iota R_\lambda)\uparrow_{R_\lambda}^{S_n} \cong K[S_n]a_\lambda$ and

Ind $(\epsilon C_\lambda)\uparrow_{C_\lambda}^{S_n} \cong K[S_n]b_\lambda$. Writing as direct sums of *minimal left ideals* of $K[S_n]$, we have $K[S_n]a_\lambda = \sum_{i=0}^r I_i$ and $K[S_n]b_\lambda = \sum_{j=0}^s J_j$. By (5.9.1) above, an I_i is S_n–equivalent to a J_j for exactly one pair, say $(i,j) = (0,0)$, i.e., $I_0 \cong J_0$ and $I_i \not\cong J_j$ for all $i,j \geq 1$. But then, by Ex.(2.9.5) above, we have $I_0 J_0 = J_0$ and $I_i J_j = (0)$ for all $(i,j) \neq (0,0)$. Now we have $K[S_n]c_\lambda = K[S_n]a_\lambda b_\lambda \subseteq (K[S_n]a_\lambda)(K[S_n]b_\lambda) = J_0$ which means that $K[S_n]c_\lambda = J_0$ (both being comparable minimal left ideals of $K[S_n]$). Thus we get that $K[S_n]c_\lambda = J_0 = [\lambda]$, as required. $\Diamond$

5.9.5 Remark: Without directly using the facts about the Frobenius-Young modules V_λ, it is possible to prove that for $\lambda \neq \mu$, $[\lambda]$ is *not equivalent to* $[\mu]$. We leave the details as an exercise. This would then yield an independent third method of constructing the family $\mathrm{Irr}_K(S_n) = \{[\lambda] \mid \lambda \vdash n\}$. ∎

5.10 Exercises

Representations considered are of a finite group G and finite dimensional too, over an algebraically closed field K such that $K[G]$ is semi–simple.

1. A group G is said to be *ambivalent* if each element is conjugate to its inverse. Show that **(i)** an abelian group is ambivalent if and only if every element is equal to its inverse, **(ii)** the complex character table of an ambivalent group is real valued and **(iii)** S_n is ambivalent for *all n* but A_4 is *not* ambivalent.

2. Show that the formal infinite product $\prod_{n=1}^\infty (1 - x^n)^{-1}$ is a *generating function* for the *partition function* $p(n)$, i.e.,

$$\prod_{n=1}^\infty (1 - x^n)^{-1} = 1 + x + 2x^2 + \cdots + p(n)x^n + \cdots$$

   Verify that $p(20) = 627$.

3. Let r be the number of non–zero parts of the *cycle type* of an element $\sigma \in S_n$. Show that $\sigma \in A_n \iff n - r$ is even.

4. Find c_λ (5.5.5) and square it for $T_\lambda = \begin{array}{ccc} 1 & 4 & 6 \\ 2 & 5 \\ 3 \end{array}$

5. Show that $(257)(34)(16) \notin R_\lambda \cdot C_\lambda$ (5.5.3) for $T_\lambda = \begin{matrix} 1 & 4 & 6 \\ 2 & 5 & \\ 3 & 8 & \\ 7 & & \end{matrix}$

6. Show that the conjugacy classes and the character table of S_3 are given as follows.
$C_1 = \{1\},\, C_2 = \{(12),(13),(23)\}$ and $C_3 = \{(123),(132)\}$.

	C_1	C_2	C_3
χ_1	1	1	1
χ_2	1	-1	1
χ_3	2	0	-1

Show also that the two dimensional irreducible representation of S_3 is given by $(12) \mapsto \begin{pmatrix} -1 & -1 \\ 0 & 1 \end{pmatrix}$ and $(23) \mapsto \begin{pmatrix} 0 & 1 \\ 1 & 0 \end{pmatrix}$.

7. Show that the conjugacy classes and the character table of S_4 are given as follows. $C_1 = \{1\}$, $C_2 = \{(12),(13),(14),(23),(24),(34)\}$, $C_3 = \{(12)(34),(13)(24),(14)(23)\}$, $C_4 = \{(123),(132),(124),(142),(134),(143),(234),(243)\}$ and $C_5 = \{(1234),(1243),(1324),(1342),(2134),(2143)\}$.

	C_1	C_2	C_3	C_4	C_5
χ_1	1	1	1	1	1
χ_2	1	-1	1	1	-1
χ_3	2	0	2	-1	0
χ_4	3	1	-1	0	-1
χ_5	3	-1	-1	0	1

8. Show that the conjugacy classes of S_5 are given as follows (wherein $h_i = |C_i|$, $1 \le i \le h$): $C_1 = \{1\}$, $C_2 = \{(12),\cdots,(45)\}$ with $h_2 = 10$, $C_3 = \{(12)(34),\cdots,(25)(34)\}$ with $h_3 = 15$, $C_4 = \{(123),\cdots,(354)\}$ with $h_4 = 20$, $C_5 = \{(123)(45),\cdots,(345)(12)\}$ with $h_5 = 20$, $C_6 = \{(1234),\cdots,(5432)\}$ with $h_6 = 30$ and $C_7 = \{(12345),\cdots,(54321)\}$ with $h_7 = 24$.
Write down the character table of S_5 and verify that it is an integral matrix. (Use (4.1.9) above, to derive the characters of S_5 from those of $S_4 \subseteq S_5$.)

9. Let (V,ϑ) be the *permutation representation* of S_n for the natural action of S_n on the set $\{1,\cdots,n\}$, i.e., V is of dimension n with a basis

$\{e_1, \cdots, e_n\}$ and $\vartheta(\sigma)(\sum_{i=1}^{n} a_i e_i) = \sum_{i=1}^{n} a_i e_{\sigma(i)}$. Show that the Specht module $W_{(n-1,1)}$ is isomorphic to the hyperplane $\{\sum_{i=1}^{n} a_i e_i \mid \sum_{i=1}^{n} a_i = 0\}$ in V. Furthermore, show that the character $\chi_{(n-1,1)}$ is given by

$$\chi_{(n-1,1)}(\sigma) = |\{j \mid \sigma(j) = j, 1 \leq j \leq n\}|, \ \forall \ \sigma \in S_n.$$

10. Let V be as above and $V^{\otimes n}$ be the n–fold tensor power of V which is an S_n–module under the natural action $\sigma(v_1 \otimes \cdots \otimes v_n) = v_{\sigma(1)} \otimes \cdots \otimes v_{\sigma(n)}$. Show that $S_K(m, n) \approx \mathrm{End}_{S_n}(V^{\otimes n})$, as K–algebras and hence or otherwise deduce that $S_K(m, n)$ is semi–simple (for all m and n) whenever $K[S_n]$ is semi–simple.

See [24] and [25] for proofs of the following three results.

11. **Determinantal formula:** Let $\lambda = (\lambda_1, \cdots, \lambda_r) \vdash n$ and $d_\lambda = n!/h_\lambda$ be the *Hook–length formula* (5.8.5). Show that the following identity holds, namely,

$$\frac{n!}{h_\lambda} = d_\lambda = n! \det\left(\frac{1}{(\lambda_i - i + j)!}\right), \ 1 \leq i, j \leq r,$$

(with the usual convention that $0! = 1$ and $r! = 0 = 1/r!$ for $r < 0$). This is a formula which uses only the parts of the partition λ and is called the *determinantal formula* for the dimensions of irreducible representations of S_n. For instance, we have

$$\frac{6!}{5 \cdot 3^2} = d_{(3,2,1)} = 6! \begin{vmatrix} \frac{1}{3!} & \frac{1}{4!} & \frac{1}{5!} \\ \frac{1}{1!} & \frac{1}{2!} & \frac{1}{3!} \\ \frac{1}{(-1)!} & \frac{1}{0!} & \frac{1}{1!} \end{vmatrix} = 16.$$

12. Show that the determinant of the character table of S_n, as an integral $p(n) \times p(n)$ matrix, is equal to the product of all the parts of all partitions of n, i.e.,

$$\det\left(\chi_\lambda(C_\mu)\right) = \prod_{(\lambda_1, \cdots, \lambda_r) \vdash n} (\lambda_1 \cdots \lambda_r).$$

13. **Branching Theorem:** Let $S_{n-1} \subseteq S_n \subseteq S_{n+1}$ in a natural way. Given $\lambda = (\lambda_1, \cdots, \lambda_r) \vdash n$, let $\lambda^{j\pm} = (\lambda_1, \cdots, \lambda_j \pm 1, \cdots, \lambda_r) \vdash n \pm 1$ according as $\lambda_{j-1} > \lambda_j$ or $\lambda_j > \lambda_{j+1}$. Then we have

$$[\lambda]\downarrow^{S_n}_{S_{n-1}} = \bigoplus_{\substack{j \\ \lambda_j > \lambda_{j+1}}} [\lambda^{j-}] \ \text{ and } \ [\lambda]\uparrow^{S_{n+1}}_{S_n} = \bigoplus_{\substack{j \\ \lambda_{j-1} > \lambda_j}} [\lambda^{j+}].$$

For instance, we have

(i) $[3,2,1] \downarrow_{S_5}^{S_6} = [2^2,1] \oplus [3,1^2] \oplus [3,2]$　while

(ii) $[3,2,1] \uparrow_{S_6}^{S_7} = [4,2,1] \oplus [3^2,1] \oplus [3,2^2] \oplus [3,2,1^2]$.

14. **Theorem:** The group S_n is characterised by its character table, i.e., if a finite group G has the same character table as that of S_n, then G is isomorphic to S_n.

See [19], [13], [2], etc., for the following and its generalisations.

15. **Schur algebras:** Let (X_{ij}) be an $m \times m$ matrix of indeterminates over K. For a positive integer $n \leq m$, let $I(m,n) = \{\underline{j} = (j_1, \cdots, j_n) \in \mathbb{N} \mid 1 \leq j_i \leq m, \forall i\}$ be the set of n-tuples of positive integers not exceeding m. We have a natural action of S_n on $I(m,n)$, namely, $\underline{j}^\sigma = (j_{\sigma(1)}, \cdots, j_{\sigma(n)})$.

Let $A_K(m,n)$ be the space of all homogeneous polynomials in the X_{ij}'s of total degree n. For given $\underline{p}, \underline{q} \in I(m,n)$, let $X_{\underline{p},\underline{q}} = X_{p_1 q_1} \cdots X_{p_n q_n}$ be the associated monomial of degree n. Show that

(i) $X_{\underline{p},\underline{q}} = X_{\underline{j},\underline{k}} \iff \underline{p}^\sigma = \underline{j}$ and $\underline{q}^\sigma = \underline{k}$ for some $\sigma \in S_n$,

(ii) $X_{\underline{p},\underline{q}}$'s form a basis of $A_K(m,n)$ and

(iii) $\dim_K(A_K(m,n)) = \binom{m^2+n-1}{n}$.

Let $S_K(m,n) = \mathrm{Hom}_K(A_K(m,n), K)$ be the *dual* of $A_K(m,n)$. Show that the vector space $S_K(m,n)$ is an associative (but non-commutative) algebra, called the *Schur algebra*, under the multiplication defined by

$$(\xi\eta)(X_{\underline{p},\underline{q}}) = \sum_{\underline{j} \in I(m,n)} \xi(X_{\underline{p},\underline{j}})\eta(X_{\underline{j},\underline{q}}), \ \forall\, \xi, \eta \in S_K(m,n) \ \& \ \underline{p},\underline{q} \in I(m,n)$$

The multiplicative identity is the element $\varepsilon \in S_K(m,n)$ where $\varepsilon(X_{\underline{p},\underline{q}}) = \delta_{\underline{p},\underline{q}}$ for all $\underline{p}, \underline{q} \in I(m,n)$. ■

Chapter 6

Representations of the Alternating Group A_n

In this chapter, we shall use the techniques developed in Chapter 4 and the representations of S_n studied in Chapter 5 to determine all irreducible representations of the Alternating group A_n over an algebraically closed field K of characteristic 0. (Cf. [25], [48], etc.)

6.1 Conjugacy Classes of A_n

Let $\sigma \in A_n$ be of cycle type $\alpha \vdash n$. Since A_n is of index 2 in S_n, the conjugacy class C_α of S_n remains one or splits into a union $C_\alpha^+ \cup C_\alpha^-$ of two conjugacy classes of A_n (in the latter case both are of equal cardinality) (4.6.1). Now we shall characterise these two possibilities in terms of the parts of the partition α.

6.1.1 Split partitions: A partition $\alpha \vdash n$ such that the conjugacy class C_α of S_n splits into two classes $C_\alpha^\pm$ in A_n is called a *split partition* of n. (Note that such an α is the cycle type of an element in A_n.)

The set of all split partitions of n is denoted by $SP(n)$. The following is a simple characterisation of the subset $SP(n)$ of split partitions in $P(n)$.

6.1.2 Proposition: *A partition $\alpha \vdash n$ is a split partition if and only if all the parts of α are odd and distinct.*

Proof: We have to prove that the conjugacy class C_α of S_n splits into a union of conjugacy classes $C_\alpha^\pm$ of A_n if and only if all the parts of α are odd and distinct. To this end, we have only to interpret the conditions of (4.6.1) above, in terms of the parts of α. Let $\sigma \in A_n$ be of cycle type $\alpha = (\alpha_1, \cdots, \alpha_r) \vdash n$. It is then equivalent to showing that $\alpha \notin SP(n)$ if and only if $C_{S_n}(\sigma)$ contains an odd permutation.

If a part of α is even, (say, α_i is even), then σ commutes with its α_i–cycle which is an odd permutation. On the other hand, suppose $\sigma = \sigma_1 \cdots \sigma_r$ is the cycle decomposition of σ with $\sigma_i = (a_1, \cdots, a_k)$ and $\sigma_{i+1} = (b_1, \cdots, b_k)$ being both k–cycles, k odd. Let $\tau = (a_1, b_1) \cdots (a_k, b_k)$ which is an odd permutation. It is clear that τ commutes with σ.

Conversely, suppose that all the parts of α are *odd and distinct.* Then we have to show that σ does not commute with any odd permutation. Let $\tau \in C_{S_n}(\sigma)$. By the uniqueness of cycle decomposition, we get that $\tau\sigma = \sigma\tau$ if and only if $\tau\sigma_i = \sigma_i\tau$ for all i, $1 \leq i \leq r$. This implies that $\sigma_i = (\tau(a_1), \cdots, \tau(a_k))$ (as cycles), i.e., $\tau(a_1) = a_p$, $\tau(a_2) = a_{p+1}$, $\cdots$, $\tau(a_k) = a_{p-1}$ for some p. Hence it follows that $\tau(a_j) = \sigma_i^{p-1}(a_j)$ for all j, $1 \leq j \leq k$, i.e., τ agrees with $\sigma_i^{p_i}$ on the symbols $a_1, \cdots, a_k$ for some p_i. Thus we get that $\tau = \prod_{i=1}^r \sigma_i^{p_i}$ which is indeed even, i.e., $\tau \in C_{A_n}(\sigma)$, as required. $\Diamond$

6.1.3 Proposition: *The set $SP(n)$ of all split partitions is naturally bijective with the set $SCP(n)$ of all self-conjugate partitions of n, i.e., $SCP(n) = \{\lambda \in P(n) \mid \lambda = \lambda'\}$.*

Proof: Given $\lambda \in SCP(n)$, let $\alpha(\lambda) = (h_{11}(\lambda), \cdots, h_{dd}(\lambda))$ where d is the length of the diagonal of the Young diagram of shape λ (5.4.1) and $h_{ii}(\lambda)$, $1 \leq i \leq d$, is the hook–length of its ith diagonal position (5.8.2). In fact, we have $h_{ii}(\lambda) = 2(\lambda_i - i) + 1$ for all i, $1 \leq i \leq d$. It is easy to check that the map $\lambda \mapsto \alpha(\lambda)$ is a bijection of $SCP(n)$ onto $SP(n)$, as required. $\Diamond$

6.1.4 Remark: Let $h(G)$ be the number of distinct conjugacy classes of a finite group G. Then $h(A_n)$ is given by

$$\begin{aligned} h(A_n) &= \text{Card}(SEP(n)) + \text{Card}(SP(n)) \\ &= \text{Card}(SEP(n)) + \text{Card}(SCP(n)) \end{aligned}$$

where $SEP(n)$ is the set of all *even partitions* of n, i.e., cycle types of all even permutations, $SP(n)$ is the set of all split partitions and $SCP(n)$ is the set of all self–conjugate partitions. ∎

6.2 Irreducible Representations of A_n

By (4.6.5) above, all the irreducible representations of A_n are known once we know how to classify the partitions $\lambda \vdash n$ into *two cases* that $V_\lambda \downarrow_{A_n}^{S_n}$ *remains irreducible* or *splits* as an A_n–module. We are now in a position to do this with the additional information that $V_\lambda = [\lambda]$ (5.9.4). First we set the following notation.

Notation: Given an irreducible representation (V, ρ) of S_n, we write $V^0 = V\downarrow_{A_n}^{S_n}$ if the latter is irreducible for A_n and $V^+ \oplus V^- = V\downarrow_{A_n}^{S_n}$ if the latter splits (in which case it splits into two inequivalent irreducible (but conjugate) representations of A_n).

Note: Since $[\lambda]\downarrow_{A_n}^{S_n} = [\lambda']\downarrow_{A_n}^{S_n}$ (by (5.9.3) above), whenever the objects $[\lambda]^0$, $[\lambda]^\pm$ are defined for λ so are the similar ones for λ' and vice–versa. In fact, we have $[\lambda]^0 = [\lambda']^0$ and $[\lambda]^\pm = [\lambda']^\pm$ for all $\lambda \vdash n$.

The following is the main result of this chapter.

6.2.1 Theorem: *Let $\lambda \vdash n$ and λ' be the conjugate of λ. Then we have the following.*
1. *$[\lambda]^0 = [\lambda]\downarrow_{A_n}^{S_n}$ is irreducible for A_n $\iff$ $\lambda \neq \lambda'$.*
2. *$[\lambda]\downarrow_{A_n}^{S_n} = [\lambda]^+ \oplus [\lambda]^-$ $\iff$ $\lambda = \lambda'$.*
3. *The three kinds of irreducible representations $[\lambda]^0$, $[\lambda]^+$ and $[\lambda]^-$ are mutually inequivalent.*
4. *Every irreducible representation of A_n is equivalent to one of the three kinds above, i.e., $\mathrm{Irr}_K(A_n) = \{[\lambda]^0 \mid \lambda \neq \lambda'\} \cup \{[\lambda]^\pm \mid \lambda = \lambda'\}$ is a complete set of irreducible representations of A_n. In particular, the number of these representations is given by*

$$\left[\frac{1}{2} \left|P(n) - SCP(n)\right|\right] + 2\left|SCP(n)\right|$$

$$= \frac{1}{2}\left[|P(n)| + 3\left|SCP(n)\right|\right].$$

Proof: We proceed in three steps.

Step 1: Every $\lambda \vdash n$ gives rise to one or two of the irreducible S_n-modules (W_i, ϑ_i), $i = 1, 2$ and 3, as defined below.

Let (W, ϑ) be an irreducible component of $[\lambda]\downarrow_{A_n}^{S_n} = [\lambda']\downarrow_{A_n}^{S_n}$ and I_W be its inertia group (4.2.5). Recall that $I_W \supseteq A_n$ and further we have
$$I_W = S_n \quad \Longleftrightarrow \quad W \text{ is self-conjugate, i.e., } \vartheta \cong \vartheta^{(12)}, \text{ where }$$
$\vartheta^{(12)}((12)\sigma(12)) = \vartheta(\sigma)$ for all $\sigma \in A_n$.

Case I: Suppose W is self-conjugate.

Since (W, ϑ) is self-conjugate, (W, ϑ) can be extended to a representation $(W, \tilde{\vartheta})$ of S_n by defining $\tilde{\vartheta}(\sigma) = \tilde{\vartheta}(\sigma(12)) = \vartheta(\sigma)$ for all $\sigma \in A_n$. (Check this.) Since $\tilde{\vartheta} \downarrow_{A_n} = \vartheta$ and $(W, \tilde{\vartheta})$ is irreducible, we get the following two irreducible inequivalent representations of S_n, namely,
(i) $(W_1, \vartheta_1) = (W, \tilde{\vartheta}) \otimes (\iota S_n) = (W, \tilde{\vartheta})$ and
(ii) $(W_2, \vartheta_2) = (W, \tilde{\vartheta}) \otimes (\epsilon S_n)$.
It is crucial (but trivial) to note that $W_1 \not\cong W_2$.

Claim 1: We have $\text{Ind}(W)\uparrow_{A_n}^{S_n} \cong W_1 \oplus W_2$.

For $i = 1, 2$, we have
$$\left\langle \chi_{\text{Ind}(W)\uparrow_{A_n}^{S_n}}, \chi_{W_i} \right\rangle_{S_n} = \left\langle \chi_W, \chi_{W_i\downarrow_{A_n}} \right\rangle_{A_n} = \left\langle \chi_W, \chi_W \right\rangle_{A_n} = 1.$$

Thus both W_1 and W_2 occur in $\text{Ind}(W)\uparrow_{A_n}^{S_n}$ and then for dimension reasons, Claim 1 follows.

Claim 2: We have $[\lambda] \cong W_1$ or W_2.

Using Claim 1, we have
$$\left\langle \chi_{(W_1 \oplus W_2)}, \chi_{[\lambda]} \right\rangle_{S_n} = \left\langle \chi_{\text{Ind}(W)\uparrow_{A_n}^{S_n}}, \chi_{[\lambda]} \right\rangle_{S_n} = \left\langle \chi_W, \chi_{[\lambda]\downarrow_{A_n}} \right\rangle_{A_n}.$$
$$= \left\langle \chi_W, \chi_W \right\rangle_{A_n} + \cdots \geq 1.$$

Thus $[\lambda]$ ocuurs in $W_1 \oplus W_2$ and hence Claim 2 follows since $[\lambda]$ and the W_i's are all irreducible.

Case II: Suppose W is <u>not</u> self-conjugate.

We have $[\lambda]{\downarrow}_{A_n}^{S_n} = (W, \vartheta) \oplus (W, \vartheta^{(12)})$ and $I_W = A_n$ and hence by (4.5.8) above, (W, ϑ) or its conjugate $(W, \vartheta^{(12)})$ gives rise to the irreducible representation of S_n, namely,
$$(W_3, \vartheta_3) = \text{Ind}\,(\vartheta){\uparrow}_{A_n}^{S_n} = \text{Ind}\left(\vartheta^{(12)}\right){\uparrow}_{A_n}^{S_n}.$$

Claim 3: We have $[\lambda] \cong W_3$.

We have $[\lambda] \downarrow_{A_n} = (W, \vartheta) \oplus (W, \vartheta^{(12)}) = W_3 \downarrow_{A_n}$ and hence

$$\Big\langle \chi_{W_3} \, , \, \chi_{[\lambda]} \Big\rangle_{S_n} = \Big\langle \chi_{\text{Ind}(W){\uparrow}_{A_n}^{S_n}} \, , \, \chi_{[\lambda]} \Big\rangle_{S_n} = \Big\langle \chi_W \, , \, \chi_{[\lambda]\downarrow_{A_n}} \Big\rangle_{A_n}$$

$$= \Big\langle \chi_W \, , \, \chi_{W \oplus W^{(12)}} \Big\rangle_{A_n} = 1 + 0 = 1, \quad \text{as required.}$$

By Claims 2 and 3 above, we have shown the following.

Step 2: Every irreducible representation $[\lambda]$ of S_n is equivalent to (W_j, ϑ_j) for some $j = 1, 2$ or 3.

In view of (4.6.5) and (5.9.4) above, the proof of the theorem would be complete if we show the following.

Step 3: (i) $[\lambda] \cong W_1$ or W_2 $\iff$ (W, ϑ) is self–conjugate $\iff$ $\lambda \neq \lambda'$, or equivalently,
(ii) $[\lambda] \cong W_3$ $\iff$ (W, ϑ) is $\underline{\text{not}}$ self–conjugate $\iff$ $\lambda = \lambda'$.

This is immediate because of the symmetry of the situation that whatever is done above remains identical if λ is replaced by λ'. Suppose (W, ϑ) is self–conjugate. By Claim 2, we get that $[\lambda']$ is also an irreducible component of $W_1 \oplus W_2$. Now suppose $\lambda = \lambda'$. Then we have $[\lambda] = [\lambda'] = W_1$, say. But then we have

$$W_1 = [\lambda'] = [\lambda] \otimes (\epsilon S_n) = W_1 \otimes (\epsilon S_n) = W_2,$$

a contradiction since $W_1 \not\cong W_2$. Claim 3 implies the converse. $\diamond$

6.2.2 Remark: Comparing the number of conjugacy classes (6.1.4) and the number of mutually inequivalent irreducible representations (6.2.1) of A_n, we get the following *bizarre identity.*

$$h(A_n) = \text{Card}(SEP(n)) + \text{Card}(SCP(n))$$

$$= \frac{1}{2} \Big[\mathrm{Card}(P(n)) + 3\, \mathrm{Card}(SCP(n)) \Big]$$

$$= \frac{1}{2} \Big[h(S_n) + 3\, \mathrm{Card}(SCP(n)) \Big]. \qquad \blacksquare$$

6.3   A Comparison

Let us review the information we have gathered about the groups S_n and A_n and then have a comparative look at the same.

6.3.1   The group S_n

1. The conjugacy classes of S_n are naturally parametrised by the set $P(n) = \{\lambda \mid \lambda \vdash n\}$ of all partitions of n (§5.2).

2. To each $\lambda \in P(n)$, we have constructed the irreducible representation $V_\lambda = [\lambda] = K[S_n]c_\lambda$ of S_n (§§5.5 and 5.9). (This is realised as a minimal left ideal of $K[S_n]$ in a natural way.)

3. The family $\mathrm{Irr}_K(S_n) = \{V_\lambda \mid \lambda \in P(n)\}$ is a complete set of mutually inequivalent (ordinary) irreducible representations of S_n (5.5.7).

4. The minimal left ideal $K[S_n]c_\lambda$ of $K[S_n]$ has a K–basis given by (5.7.14), namely, $X_\lambda = \{\sigma c_\lambda \mid \sigma \in S_n \text{ with } T_\lambda(\sigma) \text{ is } standard\}$.

5. The irreducible characters χ_{V_λ}'s (and hence also any character χ_V) are integer valued (5.7.12).

6. The dimension d_λ of V_λ is given by the Hook–length formula (5.8.5) that $d_\lambda = n!/h_\lambda$.

6.3.2   The group A_n

In contrast, notice how indirect are the corresponding statements for this group.

1. The conjugacy classes of A_n are described in §6.1. However, a description of the set to parametrise the same does not appear to be natural in anyway.

2. A complete set of mutually inequivalent irreducible representations of A_n is given by (6.2.1)

$$\mathrm{Irr}_K(A_n) := \{[\lambda]^0 = [\lambda']^0 \mid \lambda \neq \lambda'\} \bigcup \{[\lambda]^\pm \mid \lambda = \lambda'\}.$$

However, there appears no natural way of

(i) associating each member with a conjugacy class,

(ii) nor is there an obvious way to identify each member with a minimal left ideal of $K[A_n]$.

(iii) In whatever way these could be done, it is natural to expect that the *split* conjugacy classes $C_\alpha^\pm$, corresponding to $\alpha \in SP(n)$ (i.e., all parts of $\alpha \vdash n$ are odd and distinct), should correspond to the irreducibles $[\lambda]^\pm$ associated to $\lambda \in SCP(n)$ (i.e., $\lambda = \lambda'$).

(iv) In other words, the hold seems to be on the entire collection of the irreducibles rather than the individual members.

(v) A direct combinatorial justification for the identity in (6.2.2) above, should be interesting.

3. While we have

(i) $\dim_K([\lambda]^0) = d_\lambda$ for $\lambda \neq \lambda'$ and

(ii) $\dim_K([\lambda]^+) = \dim_K([\lambda]^-) = d_\lambda/2$ for $\lambda = \lambda'$,

there appears no natural way to pick bases for $[\lambda]^\pm$, nor is there a way to locate the summands $[\lambda]^\pm$ of $[\lambda]$. In fact, it is *not clear* if the "standard" basis X_λ of $[\lambda]$ as above, *splits at all* into a union of bases for the subspaces $[\lambda]^\pm$ when $\lambda = \lambda'$.

4. While the character $\chi_{[\lambda]^0}$ (for $\lambda \neq \lambda'$) is integer valued, the characters $\chi_{[\lambda]^\pm}$ are, in general, *not* integer valued (Exs. 1, 5, 7 below).

5. Knowing all that we have for S_n, much of the struggle for the case of A_n could be avoided, at any rate drastically reduced, if one can prove <u>directly</u> that $V_\lambda \downarrow_{A_n}^{S_n}$ is irreducible if and only if $\lambda \neq \lambda'$ (or, the other equivalent version). The rest would be taken care of by the Clifford and Mackey theorems. ∎

6.4 Exercises

Representations considered are over the field $K = \mathbb{C}$.

1. Show that $[2,1]^\pm$ are the non–trivial characters of A_3 and neither is integer valued.

2. Show that the conjugacy classes of A_4 are given by
   $C_1 = \{1\}$, $C_2 = \{(12)(34),(13)(24),(14)(23)\}$,
   $C_3 = \{(123),(214),(341),(432)\}$ and $C_4 = \{(132),(241),(314),(423)\}$.

3. Show that (i) $H = \{(1),(12)(34),(13)(24),(14)(23)\}$ is the commutator subgroup of A_4 and normal in S_4,
(ii) S_4 is the semi–direct product $H \bullet S_3$ (Ex.(4.8.11) above) and
(iii) the character table of H is given by

	(1)	(12)(34)	(13)(24)	(14)(23)
χ_1	1	1	1	1
χ_2	1	−1	1	−1
χ_3	1	−1	−1	1
χ_4	1	1	−1	−1

4. With notation as above, decompose the A_4–representations $\theta_j = (\chi_j)\!\uparrow_H^{A_4}$ $(1 \leq j \leq 4)$ and show that θ_4 is irreducible. However, show that none of the S_4–representations $\vartheta_j = (\chi_j)\!\uparrow_H^{S_4}$ $(1 \leq j \leq 4)$ is irreducible.

5. Show that the character table of A_4 is given by

	C_1	C_2	C_3	C_4
χ_1	1	1	1	1
χ_2	1	1	ω	ω^2
χ_3	1	1	ω^2	ω
χ_4	3	−1	0	0

where $\omega \neq 1$ is a cube root of 1. Verify that the non–trivial characters are $[2^2]^\pm$ and that χ_4 is the character of θ_4 above.

6. Show that the conjugacy classes of A_5 are given as follows (wherein $h_i = |C_i|, 1 \leq i \leq 5$): $C_1 = \{1\}$, $C_2 = \{(12)(34),\cdots,(25)(34)\}$ with $h_2 = 15$, $C_3 = \{(123),\cdots,(354)\}$ with $h_3 = 20$, $C_4 = \{(12345),\cdots,(54321)\}$ with $h_4 = 12$ and $C_5 = \{(13245),\cdots,(53421)\}$ with $h_5 = 12$.

7. With $\alpha = (1+\sqrt{5})/2$, show that the character table of A_5 is given by

	C_1	C_2	C_3	C_4	C_5
χ_1	1	1	1	1	1
χ_2	3	−1	0	α	$-\alpha^{-1}$
χ_3	3	−1	0	$-\alpha^{-1}$	α
χ_4	4	0	1	−1	−1
χ_5	5	1	−1	0	0

8. Show that an irreducible representation of A_n is of dimension 1 or at least $n - 1$ if $n \neq 5$.   ■

Part IV

Representations of the HYPEROCTAHEDRAL GROUPS B_n AND D_n

Chapter 7

Representations of the Hyperoctahedral Group B_n

In this chapter, we shall repeat the outlay of Chapter 5 for the Hyperoctahedral subgroup B_n (7.1.0) of S_{2n} under similar assumptions as for S_n. That is to determine all irreducible representations of B_n over an algebraically closed field K such that $K[B_n]$ is semi–simple. We shall see that B_n is of order $2^n n!$ (7.1.1) and hence $K[B_n]$ is semi–simple if the characteristic of K is either 0 or a prime $> n$ (in which case $K[S_{2n}]$ may not be semi–simple). As remarked in (3.7.6) above and carried out for the symmetric group S_n in Chapter 5, we need to do four things, namely, **(i)** determine the conjugacy classes of B_n, **(ii)** for each conjugacy class (λ, μ), construct an irreducible representation $V_{(\lambda,\mu)}$ in such a way that **(iii)** $V_{(\lambda,\mu)}$ is not equivalent to $V_{(\gamma,\delta)}$ for $(\lambda, \mu) \neq (\gamma, \delta)$ and **(iv)** determine the dimensions of the $V_{(\lambda,\mu)}$'s.

We present FOUR METHODS of constructing the $V_{(\lambda,u)}$'s, namely, the first is by the Wigner–Mackey method of "little groups" which uses the S_n–theory and the other three are by imitating the corresponding methods IInd, IIIrd and IVth for S_n.

I. A FORMAL deductuion from the S_n–theory by the Wigner–Mackey method of "little subgroups".

II. FROBENIUS–YOUNG method of classifying the minimal left ideals of $K[B_n]$.

III. SPECHT method of constructing simple $K[B_n]$–modules, as suitable submodules of the polynomial algebra in n varibles over K.

IV. An ABSTRACT method similar to the one used for S_n.

The formal method is by far the quickest and the reader can skip the rest of the chapter. However, the point of presenting the IInd and IIIrd methods is that the considerations are from first principles, not requiring any of the technical results from Chapter 4 and almost everything needed is readily available at hand from the S_n–theory.

For the IInd, there is no more to it than setting the notation (following the S_n–setup) and stating the results. The IIIrd is a natural extension of the case of S_n but *not* so identical. In fact, the proofs are a little more involved and so we will sketch the necessary details. We shall also give an isomorphism between the Frobenius–Young modules and the Specht modules (by means of generators and relations) so that the apparently missing details in the IInd become inconsequential. The IVth is almost identical with that of S_n. We establish also equivalence of the irreducibles constructed by all these methods.

Determination of the dimensions (hook–length formula) is very simple in this case having known it for S_n. (Cf. [1], [7], [15], [18], [37], [48] and several other papers cited in the bibliography.)

7.1   The Hyperoctahedral Group B_n

Now we define the hyperoctahedral group B_n and prove some of its basic properties needed in the sequel. (We shall follow the same pattern as for S_n to achieve our goal.)

It is more convenient to treat S_{2n} as the group of permutations of the $2n$ symbols $\pm 1, \cdots, \pm n$ instead of $1, \cdots, 2n$.

7.1.0   The group B_n: For an integer $n \geq 2$, the *hyperoctahedral group of type B_n* or simply the *hyperoctahedral group B_n* (of rank n) is defined to be the following subgroup of S_{2n}, namely,

$$B_n = \{\, \theta \in S_{2n} \mid \theta(i) + \theta(-i) = 0, \ \forall \, i, \ 1 \leq i \leq n \}.$$

In case S_{2n} is treated as the group of permutations of $1, \cdots, 2n$, then

$$B_n = \{\, \theta \in S_{2n} \mid \theta(i) + \theta(2n - i + 1) = 2n + 1, \ \forall\, i, \ 1 \le i \le 2n\}.$$

7.1.1 The order of B_n: Any element θ of B_n is of the form

$$\theta = \begin{pmatrix} 1 & \cdots & n & -n & \cdots & -1 \\ \epsilon_1 \sigma(1) & \cdots & \epsilon_n \sigma(n) & -\epsilon_n \sigma(n) & \cdots & -\epsilon_1 \sigma(1) \end{pmatrix}$$

for *unique* $(\epsilon_1, \cdots, \epsilon_n) \in N = C_2^n$ and *unique* $\sigma \in S_n$, where $C_2 = \{\pm 1\}$ is the cyclic group of order 2 and S_n is the symmetric group on $1, \cdots, n$. In fact, B_n is the *semi-direct product* of S_n by N (Ex.(7.10.1) below). Consequently, the order of B_n is $2^n n!$.

7.1.2 Positive and Negative cycles: An element in B_n which is
(i)   a product of two ℓ–cycles in S_{2n} of the form $\theta = (a_1, \cdots, a_\ell)(-a_1, \cdots, -a_\ell)$ is called a *positive ℓ-cycle*,
(ii)   a 2ℓ–cycle in S_{2n} of the form $\theta = (a_1, \cdots, a_\ell, -a_1, \cdots, -a_\ell)$ is called a *negative ℓ-cycle*.
(iii)   a positive 2–cycle $(a, b)(-a, -b)$ is called a *positive transposition*,
(iv)   a positive 2–cycle of the form $s_j = (j, j+1)(-j, -j-1)$ is called a *simple positive transposition*, $1 \le j \le n - 1$,
(v)   a negative 1–cycle $(a, -a)$ is called a *negative transposition* or a *sign change* and
(vi)   the negative 1–cycle $s_n = (n, -n)$ is called the *simple negative transposition* .

It is clear that the order of a positive ℓ–cycle is ℓ whereas that of a negative ℓ–cycle is 2ℓ.

7.1.3   Every positive cycle in B_n is a product of positive transpositions and the simple positive transpositions generate a subgroup isomorphic to S_n, called the *symmetric part* of B_n. Likewise, every negative cycle is a product of negative transpositions and the negative transpositions generate a *normal* subgroup isomorphic to $N = C_2^n$, called the *sign change part* of B_n.

7.1.4 Proposition: *The group B_n is generated by the simple transpositios $\{s_1, \cdots, s_n\}$ satisfying the following relations:*

$$s_1^2 = s_2^2 = \cdots = s_n^2 = 1,$$

$$(s_1 s_2)^3 = (s_2 s_3)^3 = \cdots = (s_{n-2} s_{n-1})^3 \;\; = \;\; 1,$$

$$(s_{n-1} s_n)^4 = 1 \;\; \text{and} \;\; s_j \, s_k = s_k \, s_j, \; \forall \, j \text{ and } k \; \ne^{\cdot} \; j \pm 1.$$

Proof: A straightforward verification shows that the stated relations are satisfied. Secondly, any positive cycle can be easily seen to be a product of the positive transpositions $s_j, 1 \le j \le n-1$, just as one proves that any permutation on $1, \cdots, n$ is a product of the (simple) transpositions $\{(j, j+1) \mid 1 \le j \le n-1\}$ in S_n.

Lastly, a negative transposition is a product of suitable positive transpositions and the simple negative transposition s_n, namely,

$$(j, -j) = (n, j)(-n, -j)(n, -n)(n, j)(-n, -j), \; \forall \, j, \; 1 \le j \le n-1$$

and hence the result follows. $\qquad\qquad\qquad\qquad\qquad\qquad\qquad \Diamond$

7.1.5 Proposition: *The index of $B_n^{(1)}$ in B_n is* 4.

Proof: Let $B_n^{\mathrm{ab}} = B_n / B_n^{(1)}$ be the abelian quotient. For $\sigma \in B_n$, let $\bar\sigma = \sigma B_n^{(1)}$. By (7.1.4) above, the group B_n^{ab} is generated by $\{\bar s_1, \cdots, \bar s_n\}$ satisfying the following relations.

$$\bar s_1^{\,2} = \bar s_2^{\,2} = \cdots = \bar s_n^{\,2} \;\; = \;\; 1,$$

$$(\bar s_1 \bar s_2)^3 = (\bar s_2 \bar s_3)^3 = \cdots = (\bar s_{n-2} \bar s_{n-1})^3 \;\; = \;\; 1,$$

$$(\bar s_{n-1} \bar s_n)^4 = 1 \;\; \text{and} \;\; \bar s_j \, \bar s_k = \bar s_k \, \bar s_j, \; \forall \, j \;\; \text{and} \;\; k.$$

From the first two and the last relations, it follows that $\bar s_1 = \bar s_2 = \cdots = \bar s_{n-1}$ and hence we get that B_n^{ab} is generated by $\bar s_1$ and $\bar s_n$ subject to the relations that $\bar s_1^{\,2} = 1 = \bar s_n^{\,2}$ and $\bar s_1 \bar s_n = \bar s_n \bar s_1$. Thus to conclude that B_n^{ab} is of order 4, we have only to check that $\bar s_1 \ne 1$ and $\bar s_n \ne 1$.

We have $B_n^{(1)} \subseteq S_{2n}^{(1)} = A_{2n}$ and so $\bar s_n = 1 \Rightarrow s_n \in B_n^{(1)}$, i.e., $s_n = (n, -n) \in A_{2n}$ which is not true. Secondly, we have $\bar s_1 = 1 \Rightarrow s_1, \cdots, s_{n-1} \in B_n^{(1)} \Rightarrow S_n \subseteq B_n^{(1)}$ where S_n is the subgroup of B_n generated by $s_1, \cdots, s_{n-1}$. But then it follows that the non–trivial character 'sgn' on S_n must be trivial since a character of any subgroup of $B_n^{(1)}$ is trivial. This contradiction completes the proof. $\qquad \Diamond$

7.1.6 Group characters of B_n: As a corollary of (7.1.5) above, we get that the *character group* of B_n is of order 4 and in fact it is given by $\widetilde{B_n} = \{\iota, \varepsilon, \xi, \eta\}$ where $\iota, \varepsilon, \xi, \eta : B_n \longrightarrow K^*$ are defined by

$$\text{trivial} : \iota(s_j) = 1, \quad \forall\, j,\ 1 \leq j \leq n.$$

$$\text{sgn} : \varepsilon(s_j) = -1, \quad \forall\, j,\ 1 \leq j \leq n.$$

$$\text{sgn}^- : \xi(s_j) = \begin{cases} 1 & \forall\, j \leq n-1, \\ -1 & \text{for } j = n. \end{cases}$$

$$\text{sgn}^+ : \eta(s_j) = \begin{cases} -1 & \forall\, j \leq n-1, \\ 1 & \text{for } j = n. \end{cases}$$

We note that $\eta = \varepsilon\xi$. ■

7.2 The Conjugacy Classes of B_n

7.2.1 Proposition: *Every element of B_n can be uniquely expressed as a product of disjoint positive and negative cycles.*

Proof: Let $\theta \in B_n$. As an element of S_{2n}, let the cycle decomposition of θ be $\theta = \theta_1\theta_2\cdots\theta_r$. Arrange these cycles in S_{2n} such that the negative cycles of B_n, if any, occur towards the right hand side. Assume therefore that θ_j, $1 \leq j \leq s$, are not negative cycles of B_n, $s \leq r$. Fix a $j \leq s$ and let $\theta_j = (a_1, \cdots, a_\ell)$ be an ℓ–cycle in S_{2n}.

Claim: We have $A = \{a_1, \cdots, a_\ell\} \cap \{-a_1, \cdots, -a_\ell\} = \emptyset$.

For, otherwise, take an $a \in A$. We may assume that $a = a_1 = -a_k$ for some $k \leq \ell$. Since $\theta_j(a_1) = a_2$, we have $\theta_j(-a_k) = a_2$. But then

$$\theta_j(-a_1) = \theta_j(a_k) = \theta(a_k) = -\theta(-a_k) = -\theta(a_1) = -\theta_j(a_1) = -a_2.$$

Since $\theta_j(a_2) = a_3$, we get that $\theta_j(-a_2) = -a_3$ and so on. Hence

$$\theta_j = (a_1, \cdots, a_k, -a_2, -a_3, \cdots, -a_{k-1}) = (a_1, \cdots, a_{k-1}, -a_1, \cdots, -a_{k-1})$$

which means that θ_j is a negative cycle in B_n contradicting the assumption. This proves the claim.

If $\overline{\theta}_j = (-a_1, \cdots, -a_\ell)$, then $\overline{\theta}_j$ must occur in the decomposition for θ, i.e., $\overline{\theta}_j = \theta_k$ for some $k \leq s$. Thus we get that

$$\theta_1\theta_2\cdots\theta_s = \theta_1\overline{\theta}_1\cdots\theta_t\overline{\theta}_t$$

with $s = 2t$, i.e., θ is a product of positive and negative cycles in B_n. Uniqueness is obvious since it is true as an element of S_{2n}. $\qquad\qquad\Diamond$

7.2.2 Complementary partitions: Two partitions $\lambda \vdash a$ and $\mu \vdash b$ with $a, b \geq 0$ are said to be *complementary partitions* of an integer n if $a + b = n$. An ordered pair (λ, μ) of complementary partitions of n is denoted by $(\lambda, \mu) \models n$.

For instance, we have $(\lambda, (0)) \models n$, $((0), \lambda) \models n$ for all $\lambda \vdash n$ and $\big((5, 3^2, 1^3), (3, 2)\big) \models 19$, $\big((4, 2^3, 1), (7, 1)\big) \models 19$, etc.

7.2.3 Dictionary order: On the set of pairs of partitions of n, we have a natural total order '$\succeq$', called the *dictionary order*, namely, $(\lambda, \mu) \succeq (\gamma, \delta)$ if $(\lambda, \mu) = (\gamma, \delta)$, or $\lambda \succ \gamma$ or $\lambda = \gamma$ and $\mu \succeq \delta$. (This is a natural extension of the dictionary order on the set of partitions, as defined in (5.2.2) above.)

Let $\theta \in B_n$ have its cycle decomposition $\theta = \theta_1 \bar\theta_1 \theta_2 \bar\theta_2 \cdots \theta_r \bar\theta_r \vartheta_1 \vartheta_2 \cdots \vartheta_s$, where $\theta_j \bar\theta_j$ is a positive cycle of length λ_j, $1 \leq j \leq r$ and ϑ_k is a negative cycle of length μ_k, $1 \leq k \leq s$. We may assume that

$$\lambda_1 \geq \lambda_2 \geq \cdots \geq \lambda_r \quad \text{and} \quad \mu_1 \geq \mu_2 \geq \cdots \geq \mu_s.$$

Since $2(\lambda_1 + \cdots + \lambda_r + \mu_1 + \cdots + \mu_s) = 2n$, we find that $(\lambda, \mu) \models n$ where $\lambda = (\lambda_1, \cdots, \lambda_r)$ and $\mu = (\mu_1, \cdots, \mu_s)$.

7.2.4 Cycle type of an element: With notation as above, the ordered pair (λ, μ) of complementary partitions of n is called the (positive–negative) *cycle type* of $\theta \in B_n$.

It is clear that for a $\theta \in B_n$, we have $\theta \in S_n$ if and only if $\mu = 0$ and $\theta \in C_2^n$ if and only if $\lambda = 0$.

7.2.5 Theorem: *The set of conjugacy classes of B_n is naturally bijective with the set of pairs of complementary partitions of n.*

Proof: Since a conjugate of a positive (resp. negative) ℓ–cycle is again one such, it follows that the cycle type of any element in B_n (which is a pair of complementary partitions of n) is invariant under conjugation. It remains to show that any two elements $\theta, \phi \in B_n$

having the same cycle type $(\lambda, \mu) \models n$ are conjugates in B_n. (Of course, θ and ϕ are conjugates in S_{2n}.) Let the cycle decompositions of θ and ϕ in B_n be

$$\theta = \theta_1 \overline{\theta}_1 \cdots \theta_r \overline{\theta}_r \vartheta_1 \cdots \vartheta_s, \text{ and } \phi = \phi_1 \overline{\phi}_1 \cdots \phi_r \overline{\phi}_r \varphi_1 \cdots \varphi_s.$$

Let $\sigma \in S_{2n}$ be defined by

$$\sigma = \begin{pmatrix} \theta_1 & \overline{\theta}_1 & \cdots & \theta_r & \overline{\theta}_r & \vartheta_1 & \cdots & \vartheta_s \\ \phi_1 & \overline{\phi}_1 & \cdots & \phi_r & \overline{\phi}_r & \varphi_1 & \cdots & \varphi_s \end{pmatrix},$$

sending θ_j (resp. $\overline{\theta}_j$, resp. ϑ_k) elementwise to ϕ_j (resp. $\overline{\phi}_j$, resp. φ_k). It is trivial to check that $\sigma \in B_n$ and that we have

$$\sigma\theta\sigma^{-1} = \left(\prod_{j=1}^{r} \sigma\theta_j\sigma^{-1}\sigma\overline{\theta}_j\sigma^{-1}\right)\left(\prod_{k=1}^{s} \sigma\vartheta_k\sigma^{-1}\right)$$

$$= \phi_1\overline{\phi}_1 \cdots \phi_r\overline{\phi}_r\varphi_1 \cdots \varphi_s = \phi, \text{ as required.} \quad \blacksquare$$

7.3   I.   The Method of Little Groups

By Ex.(7.10.1) below, $B_n = N \bullet S_n$ is the semi–direct product of S_n by the abelian group $N = C_2^n$ where $C_2 = \{\pm 1\}$ is the cyclic group of order **2**. We can therefore apply (4.7.7) above, to get the irreducible representations of B_n in terms of those of the "little subgroups" of S_n, namely, subgroups of the form $S_\ell \times S_m$ with $\ell + m = n$.

7.3.1 Since $N = C_2^n$, we have $\widetilde{N} = (\widetilde{C_2})^n$ with $\widetilde{C_2} = \{\iota, \epsilon\}$ where ι (resp. ϵ) is the trivial (resp. non–trivial) character of C_2. The action of S_n on N or $\widetilde{N}$ is by permuting coordinates.

7.3.2 For each ordered pair of non–negative integers (ℓ, m) such that $\ell + m = n$, let $\chi_{\ell m} \in \widetilde{N}$ be defined by

$$\chi_{\ell m} = \chi_{(\ell, m)} = (\iota^\ell, \epsilon^m) = (\underbrace{\iota, \cdots, \iota}_{\ell \text{ times}}, \underbrace{\epsilon, \cdots, \epsilon}_{m \text{ times}}).$$

Obviously, the S_n–orbits in $\widetilde{N}$ are precisely $\{S_n\chi_{\ell m} \mid \ell + m = n\}$. The isotropy subgroup $H_{\ell m}$ of S_n at $\chi_{\ell m}$ is $S_\ell \times S_m$ where S (resp.

S_m) is the subgroup of S_n stabilising $\{1, \cdots, \ell\}$ (resp. $\{\ell+1, \cdots, n\}$). The inertia subgroup $I_{\ell m}$ of B_n at $\chi_{\ell m}$ is $N \bullet H_{\ell m}$. The "little groups" are therefore $\{S_\ell \times S_m \mid \ell + m = n\}$.

7.3.3 From (3.10.6) and the S_k theory, recall that we have
(i) $\text{Irr}_K(S_\ell) = \{V_\lambda \mid \lambda \vdash \ell\}$, (ii) $\text{Irr}_K(S_m) = \{V_\mu \mid \mu \vdash m\}$ and
(iii) $\text{Irr}_K(S_\ell \times S_m) = \{V_\lambda \otimes_K V_\mu \mid \lambda \vdash \ell \text{ and } \mu \vdash m\}$.

For each $(\lambda, \mu) \models n$, if $\lambda \vdash \ell$ and $\mu \vdash m$ so that $\ell + m = n$, let $U_{(\lambda,\mu)} = \left(\chi_{\ell m} \otimes (V_\lambda \otimes V_\mu)\right) \uparrow^{B_n}_{I_{\ell m}}$ (as in (4.7.6) above).

The main theorem of this section is the following.

7.3.4 Theorem : *A complete set of mutually inequivalent irreducible representations of B_n is given by* $\text{Irr}_K(B_n) = \{U_{(\lambda,\mu)} \mid (\lambda,\mu) \models n\}$.

Proof: Immediate from (4.7.7) above. $\Diamond$

7.3.5 Corollary (Hook–length formula): *For each $(\lambda, \mu) \models n$, the dimension $d_{(\lambda,\mu)}$ of the irreducible representation $U_{(\lambda,\mu)}$ is given by the hook-length formula that $d_{(\lambda,\mu)} = \dim_K(U_{(\lambda,\mu)}) = n!/h_\lambda h_\mu$ where h_λ is the hook-length of λ (5.8.4), etc.*

Proof: Let $(\lambda, \mu) \models n$ with $\lambda \vdash \ell$, $\mu \vdash m$ and $\ell + m = n$. We have

$$
\begin{aligned}
d_\lambda &= [B_n : I_{\ell m}] \left(\dim_K(V_\lambda)\right) \left(\dim_K(V_\mu)\right) \\
&= [S_n : (S_\ell \times S_m)] \frac{\ell!}{h_\lambda} \frac{m!}{h_\mu} \quad \text{(using (5.8.5))} \\
&= \frac{n!}{\ell! m!} \frac{\ell!}{h_\lambda} \frac{m!}{h_\mu} = \frac{n!}{h_\lambda h_\mu}, \quad \text{as required.} \quad \Diamond
\end{aligned}
$$

Remark: The denominator $h_\lambda h_\mu$ in the above formula is called the *hook–length* of $(\lambda, \mu) \models n$. See (7.8.1) below, for the terminology.

7.3.6 Corollary: *Let ξ be the group character of B_n as defined in (7.1.6) above. Then for all $(\lambda, \mu) \models n$, we have*
(i) $U_{(\lambda,\mu)} \otimes_K \xi \cong U_{(\mu,\lambda)}$ *and consequently,*
(ii) $U_{(\lambda,\mu)} \downarrow^{B_n}_{D_n} \cong U_{(\mu,\lambda)} \downarrow^{B_n}_{D_n}$ *(where $D_n = \text{Ker}(\xi)$).*

Proof: Immediate from (4.7.8) above, since the following are true.

1. ξ is trivial on $H = S_n$.

2. $(\xi\downarrow_N^G) = \chi_{0n} = (\varepsilon^n)$ and hence

3. $(\xi\downarrow_N^G)\chi_{\ell m} = \chi_{m\ell}$ for all ℓ and m.

4. The isotropy subgroups $H_{\ell m} = S_\ell \times S_m$ and $H_{m\ell} = S_m \times S_\ell$ of S_n (at $\chi_{\ell m}$ and $\chi_{m\ell}$) are conjugates in S_n by σ where $\sigma(j) = n - j + 1$, $1 \le j \le n$ and finally the

5. $S_m \times S_\ell = \left(S_\ell \times S_m\right)^\sigma$–module $\left(V_\lambda \otimes V_\mu\right)^\sigma$ is just $V_\mu \otimes V_\lambda$. ◊

7.3.7 Corollary: By (7.1.6) and (7.3.4) above, the group of characters of B_n is given by

$$\widetilde{B_n} = \left\{\left(U_{((0),(n))}, \iota\right), \left(U_{((1^n),(0))}, \varepsilon\right), \left(U_{((n),(0))}, \xi\right), \left(U_{((0),(1^n))}, \eta\right)\right\}. ∎$$

7.4 Young Diagrams and Tableaux

Given $(\lambda, \mu) \models n$, let $\lambda = (\lambda_1, \cdots, \lambda_r)$ and its conjugate $\lambda' = (\lambda'_1, \cdots, \lambda'_{r'})$. Likewise, let $\mu = (\mu_1, \cdots, \mu_s)$ and its conjugate $\mu' = (\mu'_1, \cdots, \mu'_{s'})$.

7.4.1 Young diagrams: Given $(\lambda, \mu) \models n$, by a *Young diagram of shape* (λ, μ) or a (λ, μ)-*diagram* $T_{(\lambda,\mu)}$ of shape (λ, μ), we mean a pair of Young diagrams T_λ and T_μ of shapes λ and μ respctively.

7.4.2 Young tableaux: Given $(\lambda, \mu) \models n$, by a *Young tableaux* or a (λ, μ)- *tableaux* $T_{(\lambda,\mu)}$ of shape (λ, μ), we mean a pair of Young tableaux T_λ and T_μ together filled with the integers $\pm 1, \cdots, \pm n$ in such a way that each i or $-i$ occurs but not both, i.e., the set of absolute values of the entries of both the tableaux together is $\{1, \cdots, n\}$.

It is clear that a (λ, μ)-tableaux is obtained by first choosing an element $(\epsilon_1, \cdots, \epsilon_n)$ in $\prod_{i=1}^n \{\pm i\}$ and then filling T_λ and T_μ with the ϵ_j's without repetition. Hence there are exactly $2^n n!$ Young tableaux of shape (λ, μ), one for each element of B_n. It is also clear that the group B_n acts on the set of all (λ, μ)-tableaux by permuting the entries ϵ_j's.

Convention: *We shall follow the same convention as in the case of S_n (5.4.3) in filling a (λ, μ)-diagram along an element $\theta \in B_n$,*

namely, fill the columns of T_λ followed by those of T_μ with $\theta(1), \cdots,$
$\theta(n)$ from top to bottom in each column.

7.4.3 Example of a (λ, μ)–tableaux: Let $\lambda = (7, 5, 2) \vdash 14$ and
$\mu = (3, 2) \vdash 5$ so that $(\lambda, \mu) \models 19$. Let $\theta \in B_{19}$ be given by

$$\theta = \begin{pmatrix} 1 & \cdots & i & \cdots & 19 & \cdots & \cdots & \cdots \\ -1 & \cdots & (-1)^i i & \cdots & -19 & \cdots & \cdots & \cdots \end{pmatrix}.$$

Then the (λ, μ)–tableaux filled along θ is the following.

$T_{(\lambda,\mu)}$:

−1	4	−7	−9	−11	−13	14

2	−5	8	10	12

−15	−17	−19

−3	6

16	18

$$T_\lambda \qquad\qquad\qquad\qquad\qquad T_\mu$$

7.4.4 Row group: Given a (λ, μ)–tableaux $T_{(\lambda,\mu)} = (T_\lambda, T_\mu)$, the
set of all elements in B_n which leave the rows of T_λ stable but those
of T_μ stable only upto sign change is a subgroup of B_n, called the *row
group* of $T_{(\lambda,\mu)}$ and denoted by $R(T_{(\lambda,\mu)})$ or simply $R_{(\lambda,\mu)}$.

It is clear that we have $R_{(\lambda,\mu)} \simeq S_{\lambda_1} \times \cdots \times S_{\lambda_r} \times B_{\mu_1} \times \cdots \times B_{\mu_s}$.

7.4.5 Column group: The *column group* of $T_{(\lambda,\mu)} = (T_\lambda, T_\mu)$ is
defined as the subgroup of B_n which leave the columns of T_μ stable but
those of T_λ stable only upto sign change and is denoted by $C(T_{(\lambda,\mu)})$
or simply $C_{(\lambda,\mu)}$.

We have $C_{(\lambda,\mu)} \simeq B_{\lambda'_1} \times \cdots \times B_{\lambda'_{r'}} \times S_{\mu'_1} \times \cdots \times S_{\mu'_{s'}}$.

As in the case of S_n, we have the following.

7.4.6 Proposition: *Given a $(\lambda, \mu) \models n$ and a (λ, μ)–tableaux $T_{(\lambda,\mu)}$,
we have the following.*
1. $R_{(\lambda,\mu)} = C_{(\mu',\lambda')}$. *Note the change in the order of λ' and μ'.*

2. $R_{(\lambda,\mu)} \cap C_{(\lambda,\mu)} = \{1\}$ *and*

3. $R(\sigma T_{(\lambda,\mu)}) = \sigma R(T_{(\lambda,\mu)})\sigma^{-1}$; $C(\sigma T_{(\lambda,\mu)}) = \sigma C(T_{(\lambda,\mu)})\sigma^{-1}, \forall \sigma \in B_n$.

Proof: Straightforward verification.   ◊

7.4.7 Young subgroups: Given a $(\lambda, \mu) \models n$, the row groups of all the (λ, μ)–tableaux are uniquely determined upto conjugacy by their shape (λ, μ), called the *Young subgroups* of shape (λ, μ) and are denoted by $R_{(\lambda,\mu)}$.   ■

7.5   II. Frobenius–Young Modules for B_n

Given a $(\lambda, \mu) \models n$, we define a minimal left ideal $V_{(\lambda,\mu)}$ in $K[B_n]$ and show that the family $\{V_{(\lambda,\mu)}\}$ is a complete set of mutually non–isomorphic simple modules, called the *Frobenius-Young modules* of $K[B_n]$. The statements and their proofs being almost identical with their forerunners in §5.5 above, we omit the details.

Let T be (λ, μ)–tableaux with its row and column groups $R(T)$ and $C(T)$ respectively. Let

(i) $a_{(\lambda,\mu)} = a_{(\lambda,\mu)}(T) = \sum_{\sigma \in R(T)} \sigma$, **(ii)** $b_{(\lambda,\mu)} = b_{(\lambda,\mu)}(T) = \sum_{\tau \in C(T)} \varepsilon(\tau)\tau$

(iii) $c_{(\lambda,\mu)} = a_{(\lambda,\mu)}b_{(\lambda,\mu)} = \sum_{\sigma \in R(T), \tau \in C(T)} \varepsilon(\tau)\sigma\tau$, where ε is the 'sgn' character of B_n as defined in (7.1.6) above.

7.5.0 Examples: (i) Let $(\lambda, \mu) = ((0), (n))$ so that we have $R(T) = B_n$, $C(T) = \{ 1 \}$, $b_{((0),(n))} = 1$ and $a_{((0),(n))} = \sum_{\sigma \in B_n} \sigma = c_{((0),(n))}$. It follows that $\theta c_{((0),(n))} = c_{((0),(n))}\theta = c_{((0),(n))}$ for all $\theta \in B_n$. Consequently, $V_{((0),(n))}$ is a 1–dimensional 2–sided ideal which affords the trivial character ι of B_n.

(ii) Let $(\lambda, \mu) = ((n), (0))$ so thàt we have $R(T) = S_n$, the symmrtric part of B_n and $C(T) = C_2^n$, the sign–change part of B_n, as defined in (7.1.3) above. It follows that $V_{((n),(0))}$ is a 1–dimensional 2–sided ideal which affords the 'sgn$^-$' character ξ of B_n (7.1.6).

(iii) Let $(\lambda, \mu) = ((0), (1^n))$ so that we have $R(T) = C_2^n$ and $C(T) = S_n$. It follows that $V_{((0),(1^n))}$ is a 1–dimensional 2–sided ideal which affords the 'sgn$^+$' character η of B_n (7.1.6).

(iv) Let $(\lambda, \mu) = ((1^n), (0))$ so that $R(T) = \{1\}$ and $C(T) = B_n$. It follows that $V_{((1^n),(0))}$ is a 1–dimensional 2–sided ideal which affords the 'sgn' character ε of B_n (7.1.6).

7.5.1 Proposition: *Given a (λ, μ)–tableaux, we have the following.*
(i) $c_{(\lambda,\mu)} \neq 0$,   **(ii)** $\sigma c_{(\lambda,\mu)} = c_{(\lambda,\mu)}$ *for all $\sigma \in R_{(\lambda,\mu)}$ and*
(iii) $c_{(\lambda,\mu)}\tau = \varepsilon(\tau)c_{(\lambda,\mu)}$ *for all $\tau \in C_{(\lambda,\mu)}$.*

Proof: The first follows as in (5.5.1) above, since the row and column groups of a tableaux have no non–trivial common elements. The others are immediate from definitions.   ◊

7.5.2 Lemma (Von Neumann):   *Let $T = (T_\lambda, T_\mu)$ be a (λ, μ)–tableaux with its row and column groups $R(T)$ and $C(T)$ respectively. Then $\theta \in C(T) \cdot R(T) \iff a, b$ are in the same row of θT_λ (resp. θT_μ) $\Rightarrow c, d$ are in different columns of θT_λ (resp. θT_μ) where $c = \pm a$ and $d = \pm b$.*

Proof: Quite similar to the proof of (5.5.2) above.   Care must, however, be taken to match the argument for the pair of diagrams and the sign changes involved.   We present the full details as a test case.

Let $\theta = \tau\sigma$ with $\tau \in C(T)$ and $\sigma \in R(T)$. Suppose a and b are in the same row of $\theta T_\lambda = \tau\sigma T_\lambda = (\tau\sigma\tau^{-1})\tau T_\lambda$. Now, if c and d are in the same column of T_λ, then c and d are in the same column of τT_λ since $\tau \in C(T)$. But then, $\tau\sigma\tau^{-1}$ being a row element of τT_λ, we get that $\pm c$ and $\pm d$ are in different rows of $(\tau\sigma\tau^{-1})\tau T_\lambda = \theta T_\lambda$, contradicting the assumption that $a = \pm c$ and $b = \pm d$ are in the same row of θT_λ. Hence $c = \pm a$ and $d = \pm b$ are in different columns of θT_λ. The argument is similar for the other case too.

Conversely, suppose that a and b are in the same row of θT_λ implies that $c = \pm a$ and $d = \pm b$ are in different columns of T_λ. Then all the entries of the first column of T_λ lie in different rows of θT_λ. Hence we can find a $\phi_1 \in R(\theta T_\lambda)$ such that (upto signs) the first column of $\phi_1 \theta T_\lambda$ is filled with the entries of the first column of T_λ. Repeating this with the other columns of T_λ, we arrive at a sequence of elements $\phi_j \in R(\theta T_\lambda)$ such that (upto signs) the columns of T_λ are the same as that of $\sigma_1 \theta T_\lambda$ where $\sigma_1 = \phi_1 \cdots \phi_{\lambda_1} \in R(\theta T_\lambda)$. This means that we can find a $\tau_1 \in C(T_\lambda)$ such that $\tau_1 T_\lambda = \sigma_1 \theta T_\lambda$. We can assume that $\sigma_1 \in R(\theta T)$ by defining it to be identity on the entries of θT_μ and similarly that $\tau_1 \in C(T)$. In other words, we have $\tau_1 T = (\tau_1 T_\lambda, T_\mu)$ and $\sigma_1 \theta T = (\sigma_1 \theta T_\lambda, \theta T_\mu)$.

By the symmetry of the hypothesis, we can find $\sigma_2 \in R(\theta T)$ and $\tau_2 \in C(T)$ such that $\tau_2 T_\mu = \tau_2 \theta T_\mu$ with $\tau_2 T = (T_\lambda, \tau_2 T_\mu)$ and $\sigma_2 \theta T = (\theta T_\lambda, \sigma_2 \theta T_\mu)$. We have $\sigma_j = \theta \rho_j \theta^{-1}$ for some $\rho_j \in R(T)$, $j = 1, 2$. Since σ_1 commutes with σ_2, it is obvious that ρ_1 commutes with ρ_2. Again, by definition of τ_1 and τ_2, we find that τ_1 commutes with τ_2. Let $\sigma = \sigma_1 \sigma_2$, $\tau = \tau_1 \tau_2$ and $\rho = \rho_1 \rho_2$ so that we have $\sigma \in R(\theta T)$, $\tau \in C(T)$ and $\rho \in R(T)$ with $\tau T = (\tau_1 T \lambda, \tau_2 T_\mu) = (\sigma_1 \theta T_\lambda, \sigma_2 \theta T_\mu) = \sigma \theta T$ and hence $\tau = \sigma \theta = \theta \rho_1 \rho_2 = \theta \rho$ or $\theta = \tau \rho^{-1} \in C(T) \cdot R(T)$. ◊

7.5.3 Corollary: *If $\theta \notin C(T) \cdot R(T)$, there exists a transposition u common to $C(T)$ and $R(\theta T)$ and another transposition $v \in R(T)$ such that $u\theta = \theta v$.*

For, there exist a, b belonging to the same row of θT with $c = \pm a$ and $d = \pm b$ belonging to the same column of T. Hence $C(T) \cap R(\theta T)$ contains u, a positive transposition. But then there is a positive transposition $v \in R(T)$ such that $u = \theta v \theta^{-1}$, i.e., $\theta u = v \theta$.   ◊

7.5.4 Lemma: *Let $x \in K[B_n]$. Then $x = ac_{(\lambda,\mu)}$ for some $a \in K$ if and only if $\sigma x \tau = \varepsilon(\tau) x$ for all $\sigma \in R(T)$ and $\tau \in C(T)$.*

Proof: Similar to (5.5.4) above.   ◊

7.5.5 Theorem: *We have $c^2_{(\lambda,\mu)} = ac_{(\lambda,\mu)}$ for some $a \neq 0$ and consequently, $c_{(\lambda,\mu)} K[B_n] c_{(\lambda,\mu)} = K c_{(\lambda,\mu)}$.*

Proof: Similar to (5.5.5) and (5.5.6) above.   ◊

7.5.6 Theorem (Frobenius–Young): *Given a $(\lambda, \mu) \models n$, the left ideal $V_{(\lambda,\mu)} = K[B_n] c_{(\lambda,\mu)}$ is minimal in $K[B_n]$ and hence affords an irreducible representation of B_n, called the Frobenius-Young module associated to (λ, μ).*

Proof: Identical with (5.5.7) above, having known the counter-parts of the main ingredients like the Von Neumann lemma, etc. ◇

7.5.7 Theorem: *The family $\mathrm{Irr}_K(B_n) = \{V_{(\lambda,\mu)} \mid (\lambda,\mu) \models n\}$ is a complete set of inequivalent irreducible representations of B_n.*

Proof: Imitate the proof of (5.5.8) or (5.5.11) above. See also (7.6.7) and (7.7.9) below. ◇

7.5.8 Remark: By (7.1.6) and (7.5.0) above, the group of characters of B_n is given by

$$\widetilde{B_n} = \left\{ \left(V_{((0),(n))}, \iota\right), \left(V_{((1^n),(0))}, \varepsilon\right), \left(V_{((n),(0))}, \xi\right), \left(V_{((0),(1^n))}, \eta\right) \right\}. \quad \blacksquare$$

7.6 III. Specht Modules for B_n

In this section, we shall outline the Specht construction of the irreducible representations of B_n.

7.6.1 Let $K[X_1, \cdots, X_n]$ be the polynomial algebra over K in n variables. Define $X_{-j} = -X_j$ for all $j = 1, \cdots, n$. Now the group B_n acts linearly on the polynomial algebra by permuting and sign change of the variables, i.e., $\theta(f(X_1, \cdots, X_n)) = f(X_{\theta(1)}, \cdots, X_{\theta(n)})$ for all $\theta \in B_n$.

7.6.2 (i). For non–zero integers $a_1, \cdots, a_\ell$ between $-n$ and n, define a Vandermonde type determinant, namely,

$$\Omega(a_1, \cdots, a_\ell) = \begin{vmatrix} 1 & \cdots & \cdots & 1 \\ X_{a_1}^2 & \cdots & \cdots & X_{a_\ell}^2 \\ X_{a_1}^4 & \cdots & \cdots & X_{a_\ell}^4 \\ \vdots & \vdots & \vdots & \vdots \\ X_{a_1}^{2\ell-2} & \cdots & \cdots & X_{a_\ell}^{2\ell-2} \end{vmatrix}$$

which is the usual Vandermonde determinant with the variables replaced by their squares. Consequently, we get that

$$\Omega(a_1, \cdots, a_\ell) = \prod_{1 \leq j \leq k \leq \ell} (X_{a_k}^2 - X_{a_j}^2).$$

(ii). Given $(\lambda, \mu) \models n$, let $\lambda \vdash \ell$ and $\mu \vdash m$ with $\ell + m = n$. Let $T_{(\lambda,\mu)}(\theta) = (T_\lambda(\theta), T_\mu(\theta))$ be a Young tableaux of shape (λ, μ), filled along a $\theta \in B_n$, where $T_\lambda(\theta)$ is filled with $\{\theta(i) \mid i \leq \ell\} = \{a_{jk}\}$ and $T_\mu(\theta)$ filled with $\{\theta(j) \mid \ell + 1 \leq j \leq \ell + m = n\} = \{b_{jk}\}$; $-n \leq a_{jk}, b_{jk} \leq n$.

Let $\Delta_{(\lambda,\mu)}(\theta) = \Gamma_\lambda(\theta)\Omega_\lambda(\theta)\Omega_\mu(\theta)$ where $\Gamma_\lambda(\theta) = \prod_{j=1}^{\lambda_1'} \prod_{k=1}^{\lambda_1} X_{a_{jk}}$, $\Omega_\lambda(\theta) = \prod_{k=1}^{\lambda_1} \Omega(a_{1k}, \cdots, a_{\lambda_k'k})$ and $\Omega_\mu(\theta) = \prod_{k=1}^{\mu_1} \Omega(b_{1k}, \cdots, a_{\mu_k'k})$.

7.6.3 Specht Polynomials: Given $(\lambda, \mu) \models n$ and $\theta \in B_n$, the homogeneous polynomial $\Delta_{(\lambda,\mu)}(\theta)$, defined above, is called the *Specht polynomial* associated to θ.

We note that this polynomial is not symmetric in λ and μ because of the monomial factor $\Gamma_\lambda(\theta)$.

For the case when $\theta = 1$, for simplicity, we shall write $\Delta_{(\lambda,\mu)}$ instead of $\Delta_{(\lambda,\mu)}(1)$. It is obvious from definitions that $\Delta_{(\lambda,\mu)}(\theta) = \theta\Delta_{(\lambda,\mu)}$ for all $\theta \in B_n$ (for the action of θ on $\Delta_{(\lambda,\mu)}$ as in (7.6.1) above).

7.6.4 Examples: 1. It is quite trivial to check the following.
(i) $\Delta_{((0),(n))} = 1$, (ii) $\Delta_{((n),(0))} = \prod_{1 \leq i \leq n} X_i$,
(iii) $\Delta_{((0),(1^n))} = \prod_{1 \leq i < j \leq n}(X_j^2 - X_i^2)$ and
(iv) $\Delta_{((1^n),(0))} = (\prod_{1 \leq i \leq n} X_i)(\prod_{1 \leq i < j \leq n}(X_j^2 - X_i^2))$.
2. Let $\lambda = (5, 3, 2) \vdash 10$ and $\mu = (3, 2) \vdash 5$ so that $(\lambda, \mu) \models 15$. Let $\theta \in B_{15}$ be given by

$$\theta = \begin{pmatrix} 1 & \cdots & i & \cdots & 15 & \cdots & \cdots & \cdots \\ -1 & \cdots & (-1)^i i & \cdots & -15 & \cdots & \cdots & \cdots \end{pmatrix}.$$

Then the (λ, μ)–tableaux filled along θ is the following.

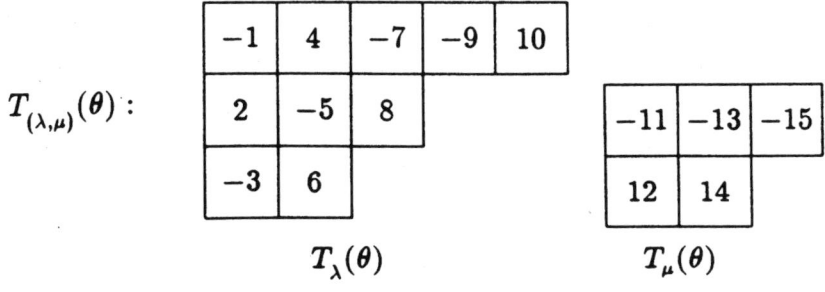

$T_{(\lambda,\mu)}(\theta):$

−1	4	−7	−9	10
2	−5	8
−3	6

$T_\lambda(\theta)$

−11	−13	−15
12	14

$T_\mu(\theta)$

We have

$$\Omega(-1,2,3) = \begin{vmatrix} 1 & 1 & 1 \\ X_{-1}^2 & X_2^2 & X_{-3}^2 \\ X_{-1}^4 & X_2^4 & X_{-3}^4 \end{vmatrix} = \begin{vmatrix} 1 & 1 & 1 \\ X_1^2 & X_2^2 & X_3^2 \\ X_1^4 & X_2^4 & X_3^4 \end{vmatrix}$$

$$= (X_3^2 - X_1^2)(X_3^2 - X_2^2)(X_2^2 - X_1^2),$$

$$\Omega(4,-5,6) = (X_6^2 - X_4^2)(X_6^2 - X_5^2)(X_5^2 - X_4^2),$$

$$\Omega(-7,8) = (X_8^2 - X_7^2) \text{ and } \Omega(-9) = 1 = \Omega(10),$$

$$\Omega(-11,12) = (X_{12}^2 - X_{11}^2) \text{ and } \Omega(-13,14) = (X_{14}^2 - X_{13}^2),$$

$$\Omega_\lambda(\theta) = \Omega(-1,2,3)\Omega(4,-5,6)\Omega(-7,8)$$

$$= (X_3^2 - X_1^2)(X_3^2 - X_2^2)(X_2^2 - X_1^2)(\cdots)(X_8^2 - X_7^2),$$

$$\Omega_\mu(\theta) = \Omega(-11,12)\Omega(-13,14) = (X_{12}^2 - X_{11}^2)(X_{14}^2 - X_{13}^2),$$

$$\Gamma_\lambda(\theta) = \prod_{i=1}^{10}(-1)^i X_i = -\prod_{i=1}^{10} X_i \text{ and}$$

$$\Delta_{(\lambda,\mu)}(\theta) = \Gamma_\lambda(\theta)\Omega_\lambda(\theta)\Omega_\mu(\theta) = -(X_1 \cdots X_{10})(\cdots)(\cdots).$$

7.6.5 Specht modules: Given $(\lambda,\mu) \models n$, the cyclic B_n–submodule of $K[X_1, \cdots, X_n]$ generated by the Specht polynomial $\Delta_{(\lambda,\mu)}$ is called the *Specht module* associated to (λ,μ) and is denoted by $W_{(\lambda,\mu)}$.

Since $\theta\Delta_{(\lambda,\mu)} = \Delta_{(\lambda,\mu)}(\theta)$, $\forall\, \theta \in B_n$, it follows that $W_{(\lambda,\mu)}$ is spanned, as a vector space over K, by $\{\Delta_{(\lambda,\mu)}(\theta) \mid \theta \in B_n\}$.

7.6.6 Theorem: *The Specht module $W_{(\lambda,\mu)}$ is simple.*

Proof: Since $K[B_n]$ is semi–simple, $W_{(\lambda,\mu)}$ is semi–simple and hence the result is equivalent to showing that $\mathrm{End}_{B_n}(W_{(\lambda,\mu)}) = K$.

Let $f \in \mathrm{End}_{B_n}(W_{(\lambda,\mu)})$. If $(\lambda,\mu) = ((0),(n))$, then $\Delta_{((0),(n))} = 1$ and the result is obvious. Let $(\lambda,\mu) \neq ((0),(n))$.

Let $T_{(\lambda,\mu)} = (T_\lambda, T_\mu)$ be the (λ,μ)–tableaux corresponding to the identity element of B_n. Let i,j be two entries in the same column of T_λ or T_μ (which exist since $(\lambda,\mu) \neq ((0),(n))$). Let $\sigma = (i,j)(-i,-j)$ be the positive transposition in B_n. Since $(X_i - X_j)$ is a factor of $\Delta_{(\lambda,\mu)}$, we have $\sigma\Delta_{(\lambda,\mu)} = -\Delta_{(\lambda,\mu)}$. Hence, we get that $\sigma f(\Delta_{(\lambda,\mu)}) = f(\sigma\Delta_{(\lambda,\mu)}) = -f(\Delta_{(\lambda,\mu)})$ which implies that $(X_i - X_j)$ divides $f(\Delta_{(\lambda,\mu)})$. Now taking $\tau = (i, -j)(-i, j)$, we conclude in a similar way that $(X_i + X_j)$ divides $f(\Delta_{(\lambda,\mu)})$ and consequently, we find that $(X_i^2 - X_j^2)$ divides $f(\Delta_{(\lambda,\mu)})$. Since $(X_i^2 - X_j^2)$ is a simple factor of Ω_λ (or Ω_μ) which is a factor of $\Delta_{(\lambda,\mu)}$, we see that $\Omega_\lambda \Omega_\mu$ divides $f(\Delta_{(\lambda,\mu)})$. On the other hand, if k is any entry in T_λ, i.e., when $\lambda \neq (0)$, look at the negative transposition $\rho = (k, -k)$ in B_n. Since X_k is a factor of Γ_λ, we have $\rho \Gamma_\lambda = -\Gamma_\lambda$ and so we get that $\rho f(\Delta_{(\lambda,\mu)}) = -f(\Delta_{(\lambda,\mu)})$. This gives that X_k divides $f(\Delta_{(\lambda,\mu)})$. This is true for all entries of T_λ and hence Γ_λ divides $f(\Delta_{(\lambda,\mu)})$. Putting all these together, we conclude that $\Delta_{(\lambda,\mu)}$ is a factor of $f(\Delta_{(\lambda,\mu)})$. But both are polynomials of the same degree and so $f(\Delta_{(\lambda,\mu)})$ is a scalar multiple of $\Delta_{(\lambda,\mu)}$. $\diamond$

7.6.7 Theorem: *The family* $\mathrm{Irr}_K(B_n) = \{W_{(\lambda,\mu)} \mid (\lambda, \mu) \vDash n\}$ *is a complete set of inequivalent irreducible representations of* B_n.

Proof: Imitating the proof of (5.6.6), we see that the annihilator ideals of $W_{(\lambda,\mu)}$ and $W_{(\gamma,\delta)}$ are *not equal* if $(\lambda, \mu) \neq (\gamma, \delta)$. $\diamond$

7.6.8 Remark: Let $S_n = \{\theta \in B_n \mid \theta(i) \geq 1, \forall i \geq 1\}$ which is the symmetric part of B_n (a subgroup, isomorphic to S_n). Given $\theta \in B_n$, let $|\theta| \in H_n$ be defined as

$$|\theta|(i) = \begin{cases} |\theta(i)|, & \forall i \geq 1, \\ -|\theta(i)|, & \forall i \leq -1. \end{cases}$$

With notation as in (7.6.3) above, we have $\Omega_\lambda(\theta) = \Omega_\lambda(|\theta|)$ and $\Omega_\mu(\theta) = \Omega_\mu(|\theta|)$ whereas $\Gamma_\lambda(\theta) = \pm\Gamma_\lambda(|\theta|)$. Consequently, we get that $\Delta_{(\lambda,\mu)}(\theta) = \pm\Delta_{(\lambda,\mu)}(|\theta|)$ and hence $W_{(\lambda,\mu)}$ is spanned by $\{\Delta_{(\lambda,\mu)}(\theta) \mid \theta \in S_n\}$ as a vector space over K.

In the next section, we shall extract a natural basis from this set of generators for $W_{(\lambda,\mu)}$. ∎

7.7 Standard Young Tableaux

Following the same pattern as in § 5.7, above, now we shall outline a proof of the basis theorem for the Specht modules for B_n, constructed in the previous section.

7.7.1 Notation: Henceforward, we keep the following notation.
1. $S_n = \{\theta \in B_n \mid \theta(i) \geq 1, \ \forall \ i \geq 1\}$, the symmetric part of B_n.
2. Given $(\lambda, \mu) \models n$, let $\lambda \vdash \ell \geq 0$ and $\mu \vdash m \geq 0$ with $\ell + m = n$.
3. Given $\theta \in S_n$, we write

$$\theta = \begin{pmatrix} 1 & \cdots & \ell & ; & \ell+1 & \cdots & \ell+m & : & \cdots & \cdots & \cdots \\ a_1 & \cdots & a_\ell & ; & b_1 & \cdots & b_m & : & \cdots & \cdots & \cdots \end{pmatrix}$$

so that $\{a_1, \cdots, a_\ell \ ; \ b_1, \cdots, b_m\}$ is a permutation of $\{1, \cdots, n\}$.
4.   Given $\theta \in S_n$, a (λ, μ)–tableaux filled along θ is $T_{(\lambda,\mu)}(\theta) = (T_\lambda(\theta), T_\mu(\theta))$ where $T_\lambda(\theta)$ is a λ–tableaux filled with $\{a_1, \cdots, a_\ell\}$ and $T_\mu(\theta)$ is a μ–tableaux filled with $\{b_1, \cdots, b_m\}$.
5. For $\theta \in S_n$, we have the following.
(i) $\Gamma_\lambda(\theta) = X_{a_1} \cdots X_{a_\ell}$,   (ii) $\Omega_\lambda(\theta) = \Omega_\lambda(X_{a_1}^2, \cdots, X_{a_\ell}^2)$,
(iii) $\Omega_\mu(\theta) = \Omega_\mu(X_{b_1}^2, \cdots, X_{b_m}^2)$ and   (iv) $\Delta_{(\lambda,\mu)}(\theta)$ is a polynomial in odd powers of X_{a_j}'s and even powers of X_{b_k}'s.

7.7.2   Standard Young tableaux: Given $\theta \in S_n$, a (λ, μ)–tableaux $T_{(\lambda,\mu)}(\theta)$ (filled along θ) is said to be *standard* if both $T_\lambda(\theta)$ and $T_\mu(\theta)$ are standard, i.e., the rows as well as columns of $T_\lambda(\theta)$ and $T_\mu(\theta)$ are strictly increasing.

We note that in a standard tableaux, the entries are all positive in both the parts.

7.7.3   Standard Specht polynomials: The Specht polynomial $\Delta_{(\lambda,\mu)}(\theta)$ associated to a standard Young tableaux $T_{(\lambda,\mu)}(\theta)$ is called a *standard Specht polynomial*.

7.7.4   Diagonal terms of Specht polynomials: The monomial

$$\mathcal{D}_{(\lambda,\mu)}(\theta) := \Gamma_\lambda(\theta) D_\lambda^2(\theta) D_\mu^2(\theta)$$

is called the *diagonal term* or the *leading term* of the Specht polynomial $\Delta_{(\lambda,\mu)}(\theta)$ where $D_\lambda(\theta)$ (resp. $D_\mu(\theta)$) is the diagonal term of the Specht polynomial $\Delta(T_\lambda(\theta))$ (resp. $\Delta(T_\mu(\theta))$), as defined in (5.7.4) above.

We note that the diagonal term $\mathcal{D}_{(\lambda,\mu)}(\theta)$ is a monomial in odd powers of X_{a_j}'s and even powers of X_{b_k}'s where the a_j's and b_k's are the entries of $T_\lambda(\theta)$ and $T_\mu(\theta)$ respectively.

Basis Theorem for Specht Modules for B_n

7.7.5 Theorem: *Given $(\lambda, \mu) \models n$, the set of standard Specht polynomials of shape (λ, μ) is a basis for the Specht module $W_{(\lambda,\mu)}$.*

Proof: We shall just imitate the proof of (5.7.5) above.

Step 1. *Linear independence of standard Specht polynomials.*

1(a): *The diagonal terms of standard Specht polynomials of shape (λ, μ) are linearly independent.*

For degree reasons, we have only to show that the diagonal terms of distinct standard tableaux are distinct. This is trivial because we have the following.
(i) $\mathcal{D}_{(\lambda,\mu)}(\theta) = \mathcal{D}_{(\lambda,\mu)}(\tau)$ implies for degree reasons (odd / even powers) that the entries of $T_\lambda(\theta)$ are a permutation of those of $T_\lambda(\tau)$ and so $\Gamma_\lambda(\theta)D_\lambda(\theta) = \Gamma_\lambda(\tau)D_\lambda(\tau)$ and $D_\mu(\theta) = D_\mu(\tau)$.
(ii) $\Gamma_\lambda(\theta)D_\lambda(\theta) = \Gamma_\lambda(\tau)D_\lambda(\tau) \Rightarrow \Gamma_\lambda(\theta) = \Gamma_\lambda(\tau)$ and $D_\lambda(\theta) = D_\lambda(\tau)$.
(iii) Consequently, we get that $T_\lambda(\theta) = T_\lambda(\tau)$ since both $T_\lambda(\theta)$ and $T_\lambda(\tau)$ are standard having the same set of entries. Similarly, **(iv)** $D_\mu(\theta) = D_\mu(\tau) \Rightarrow T_\mu(\theta) = T_\mu(\tau)$. But then **(v)** $\theta = \tau$, as required.

1(b): *Standard Specht polynomials are linearly independent.*

As in the case of S_n, we see easily that any non–trivial dependency relation among the standard Specht polynomials $\Delta_\lambda(\theta)$ gives rise to a non–trivial dependency relation among their diagonal terms $\mathcal{D}_\lambda(\theta)$. This contradicts 1(a) above.

Step 2. *Standard Specht polynomials span $W_{(\lambda,\mu)}$ as a vector space.*

Since the set of Specht polynomials $\{\Delta_{(\lambda,\mu)}(\theta) \mid \theta \in S_n\}$ span $W_{(\lambda,\mu)}$, it suffices to show that any such Specht polynomial $\Delta_{(\lambda,\mu)}(\theta)$ can be written as a linear combination of suitable *standard* Specht polynomials $\Delta_{(\lambda,\mu)}(\theta_j)$.

Suppose $T_\lambda(\theta)$ is not standard. Following the proof of Step 2 of (5.7.5) above, we write $\Delta_\lambda(\theta) = \sum_{j=1}^{\tilde{}} d_j \Delta_\lambda(\sigma_j)$ for some $\sigma_j \in S_\ell$ and

$d_j \in \mathbb{Z}^+$ where S_ℓ is the symmetric group on the entries $\{a_1, \cdots, a_\ell\}$ of $T_\lambda(\theta)$ (fixing the entries of $T_\mu(\theta)$) and for each j, $T_\lambda(\sigma_j)$ is standard. It is *very important* to note that $\Gamma_\lambda(\sigma_j) = \Gamma_\lambda(\theta)$, for all j, $(1 \leq j \leq r)$. Consequently, we get that $\Gamma_\lambda(\theta)\Omega_\lambda(\theta) = \sum_{j=1}^r d_j \Gamma_\lambda(\sigma_j)\Omega_\lambda(\sigma_j)$. Similarly, if $T_\mu(\theta)$ is not standard, we can write $\Omega_\mu(\theta) = \sum_{k=1}^s e_k \Omega_\mu(\tau_k)$, where $e_k \in \mathbb{Z}^+$, $\tau_k \in S_m$ and S_m is the symmetric group on the entries $\{b_1, \cdots, b_m\}$ of $T_\mu(\theta)$ (fixing the entries of $T_\lambda(\theta)$). Now we have

$$\Delta_{(\lambda,\mu)}(\theta) = \sum_{j=1}^r \sum_{k=1}^s d_j e_k \Gamma_\lambda(\sigma_j)\Omega_\lambda(\sigma_j)\Omega_\mu(\tau_k).$$

Let $\theta_{jk} = \sigma_j \tau_k$, $1 \leq j \leq r$, $1 \leq k \leq s$. For all j and k, we observe the following, namely,
(i) $\theta_{jk} \in S_n$, (ii) $T_{(\lambda,\mu)}(\theta_{jk}) = (T_\lambda(\sigma_j), T_\mu(\tau_k)$ is standard,
(iii) $\Gamma_\lambda(\theta_{jk}) = \Gamma_\lambda(\sigma_j)$ and (iv) $\Delta_{(\lambda,\mu)}(\theta) = \sum_{j,k} d_{jk}\Delta_{(\lambda,\mu)}(\theta_{jk})$ which is a sum of standard Specht polynomials, as required.   ◇

7.7.6 Corollary: *Representations of B_n are defined over $\mathbb{Z}$ (5.7.12) and so the character of any representation of B_n is integer valued.*

7.7.7 Remark: By similar cosiderations as above, it can be seen that the Frobenius–Young module $V_{(\lambda,\mu)}$ has a basis consisting of

$$Xc_{(\lambda,\mu)} = \left\{ \theta c_{(\lambda,\mu)} \mid \theta \in S_n, T_{(\lambda,\mu)}(\theta) \text{ is standard} \right\}.$$

Generators and Relations for Simple B_n–Modules

7.7.8 Let $(\lambda, \mu) \models n$ with $\lambda \vdash \ell$, $\mu \vdash m$ and $\ell + m = n$. Let $M_{(\lambda,\mu)}$ be the cyclic B_n–module generated by an element $f = f_{(\lambda,\mu)}$. For example, we take

$$f_{(\lambda,\mu)} = \begin{cases} c_{(\lambda,\mu)} & \text{(for Frobenius–Young module)} \\ \Delta_{(\lambda,\mu)} & \text{(for Specht module)} \end{cases}$$

The module $M_{(\lambda,\mu)}$ is subject to the following relations.

Let $\theta \in B_n$ and $T_{(\lambda,\mu)}(\theta) = (T_\lambda(\theta), T_\mu(\theta))$ be the Young tableaux filled along θ. Assume that the following are true. Let $T = T_\lambda$ or T_μ.

1. For any two entries a, b in the same column of T,

Alternacy relation: $(a, b)(-a, -b)f = -f$.

2. For any entry a in T,

Sign change relation: $(a, -a)f = \begin{cases} -f & \text{if } T = T_\lambda, \\ f & \text{if } T = T_\mu. \end{cases}$

3. Following the construction as in (5.7.9) above, let $a = a_{jk}$ be an entry in T at the $(jk)^{\text{th}}$ position and $b = b_{j(k+1)}$ in T. Let A be the set of all entries in the kth column of T below and including a. Likewise, let B be the set of all entries in the $(k+1)^{\text{th}}$ column of T above and including b. Let $C = A \cup B$. Let S_A, S_B and S_C be the subgroups of B_n generated by the positive permutations on A, B and C respectively. Let $S(A, B)$ be a complete set of coset representatives of $S_A \cdot S_B$ in S_C. Let $G(A, B) = \sum_{\sigma \in S(A,B)} \varepsilon(\sigma)\sigma$ (7.1.6), called the *Garnir element* associated to A and B.

Garnir relation: $G(A, B)f = \displaystyle\sum_{\sigma \in S(A,B)} \varepsilon(\sigma)\sigma f = 0.$

7.7.9 Remark: Proceeding by induction on (λ, μ) under the dictionary order (7.2.3) above, it can be shown that the annihilator ideal of the Frobenius–Young module $V_{(\lambda,\mu)}$ or of the Specht module $W_{(\lambda,\mu)}$ is generated by the elements in $K[B_n]$ appearing in the *alternacy, sign-change or Garnir relations*, as above. Hence $V_{(\lambda,\mu)}$ and $W_{(\lambda,\mu)}$ are isomorphic as $K[B_n]$–modules. ■

7.8 Hook–Length Formula

In this section, we shall determine the dimensions of the (ordinary) irreducible representattions of B_n. From the basis theorem (7.7.5) for the Specht modules $W_{(\lambda,\mu)}$, we have $d_{(\lambda,\mu)} = \dim_K W_{(\lambda,\mu)}$ is the number of standard Young tableaux of shape (λ, μ). It is easy to count these tableaux having at hand the hook–length formula for S_n (see also (7.3.5) above).

7.8.1 Hook–length of a diagram: The *hook-length of a diagram* of shape (λ, μ), or simply the *hook-length* of (λ, μ) is defined to be the product of the hook–lengths of its constituent diagrams of shapes λ and μ and is denoted by $h_{(\lambda,\mu)}$, i.e., $h_{(\lambda,\mu)} = h_\lambda h_\mu$.

7.8.2 Theorem (Hook–length formula): *For each* $(\lambda, \mu) \models n$, *the number* $d_{(\lambda, \mu)}$ *of standard Young tableaux of shape* (λ, μ) *is given by* $d_{(\lambda, \mu)} = n!/h_{(\lambda, \mu)}$, *which is called the* hook-length formula, *giving the dimension of the irreducible representation* $W_{(\lambda, \mu)}$ *of* B_n.

Proof: Let $(\lambda, \mu) \models n$ with $\lambda \vdash \ell$, $\mu \vdash m$ and $\ell + m = n$. All standard Young tableaux of shape (λ, μ) are obtained by first choosing any ℓ inegers $a_1, \cdots . a_\ell$ between 1 and n and then forming all possible standard tableaux of shape λ with the a_j's and of shape μ with the remaining m posive integers $b_1, \cdots, b_m$. This can be done in $\binom{n}{\ell} d_\lambda d_\mu$ ways where $d_\lambda = \ell!/h_\lambda$ (using (5.8.5) above). Thus we have

$$d_{(\lambda, \mu)} \;=\; \binom{n}{\ell} \frac{\ell!}{h_\lambda} \frac{m!}{h_\mu} = \frac{n!}{h_{(\lambda, \mu)}}, \quad \text{as required.} \qquad \Diamond$$

7.8.3 Example: The dimensions of all inequivalent irreducible representations of B_5, arranged according to the dictionary order (7.2.3), are 1,4,5,6,5,4,1; 5,15,10,5,5; 5,5,20,20,5,5; 5,20,5,5,20,5; 5,15, 10,5,5 and 1,4,5,6,5,4,1.   ∎

7.9 Irreducible Representations of B_n– An Abstract Method

In this section, we shall give a formal method of constructing the irreducible representations of B_n on similar lines as for the case of S_n. We shall also establish the equivalence of these with the previously constructed ones like the Frobenius–Young modules, for example. The method and even the details being almost identical with the earlier one, we shall omit the details except pointing out some modifications wherever needed.

Given $(\lambda, \mu) \models n$, fix a Young tableaux $T_{\lambda\mu} = T_{(\lambda,\mu)} = (T_\lambda, T_\mu)$ of shape (λ, μ) (7.4.2). Let the row and column groups of $T_{\lambda\mu}$ be given by ((7.4.4) and (7.4.5)): $R_{\lambda\mu} = R_{(\lambda,\mu)} = S_{\lambda_1} \times \cdots \times S_{\lambda_r} \times B_{\mu_1} \times \cdots \times B_{\mu_s}$ and $C_{\lambda\mu} = C_{(\lambda,\mu)} = B_{\lambda'_1} \times \cdots \times B_{\lambda'_r} \times S_{\mu'_1} \times \cdots \times S_{\mu'_{s'}}$, where $\lambda = \{\lambda_1, \cdots, \lambda_r\} \vdash \ell$ and $\mu = (\mu_1, \cdots, \mu_s) \vdash m$ with $\ell + m = n$.

Let $\iota R_{\lambda\mu}$ be the trivial character of $R_{\lambda\mu}$ and $\varepsilon C_{\lambda\mu}$ be the 'sgn' character (7.1.6) of $C_{\lambda\mu}$. Let $\mathrm{Ind}\,(\iota R_{\lambda\mu})\!\uparrow^{B_n}_{R_{\lambda\mu}}$ and $\mathrm{Ind}\,(\varepsilon C_{\lambda\mu})\!\uparrow^{B_n}_{C_{\lambda\mu}}$ be the induced representations. Now we have the following.

7.9.1 Theorem: *The representations* $\mathrm{Ind}\,(\iota R_{\lambda\mu})\!\uparrow^{B_n}_{R_{\lambda\mu}}$ *and* $\mathrm{Ind}\,(\varepsilon C_{\lambda\mu})\!\uparrow^{B_n}_{C_{\lambda\mu}}$ *of* B_n *have a* unique *irreducible subrepresentation of multiplicity* 1 *in common and it is denoted by*

$$[\lambda,\mu] = \left(\mathrm{Ind}\,(\iota R_{\lambda\mu})\!\uparrow^{B_n}_{R_{\lambda\mu}}\right) \bigcap \left(\mathrm{Ind}\,(\varepsilon C_{\lambda\mu})\!\uparrow^{B_n}_{C_{\lambda\mu}}\right).$$

Proof: Mutatis–mutandis (5.9.1) above, replacing (5.5.3) by (7.5.3) where necessary.   ◇

7.9.2 Examples: It is easy to see that we have
(i) $[(0),(n)] = \iota B_n$ and **(ii)** $[(1^n),(0)] = \varepsilon B_n$.

7.9.3 Remark: $[\mu',\lambda'] = [\lambda,\mu] \otimes (\varepsilon B_n)$ for all $(\lambda,\mu) \models n$.

Given $(\lambda,\mu) \models n$ and $(\alpha,\beta) \models n$, we shall prove that $[\lambda,\mu] \cong [\alpha,\beta]$ if and only if $(\lambda,\mu) = (\alpha,\beta)$. To do this, recall the notation (from §7.4 above) and also some facts (7.5.1) and (7.5.8), namely,
(i) $a_{\lambda\mu} = a_{(\lambda,\mu)} = \sum_{\sigma \in R_{\lambda\mu}} \sigma$, **(ii)** $b_{\lambda\mu} = b_{(\lambda,\mu)} = \sum_{\tau \in C_{\lambda\mu}} \varepsilon(\tau)\tau$,
(iii) $c_{\lambda\mu} = a_{\lambda\mu}b_{\lambda\mu} = \sum_{\sigma \in R_{\lambda\mu}\,,\,\tau \in C_{\lambda\mu}} \varepsilon(\tau)\sigma\tau$, **(iv)** $c_{\lambda\mu} \neq 0$ and
(v) $V_{(\lambda,\mu)} = K[B_n]c_{\lambda\mu}$ is a minimal left ideal of $K[B_n]$.

7.9.4 Theorem: *For all* $(\lambda,\mu) \models n$, *we have* $[\lambda,\mu] = K[B_n]c_{\lambda\mu}$ *and hence the family* $\mathrm{Irr}_K(B_n) = \{\,[\lambda,\mu] \mid (\lambda,\mu) \models n\}$ *is a complete set of irreducible representations of* B_n.

Proof: This goes verbatim with (5.9.4) above. As before, we use Ex.(4.8.8) above, to get **(i)** $\mathrm{Ind}\,(\iota R_{\lambda\mu})\!\uparrow^{B_n}_{R_{\lambda\mu}} \cong K[B_n]a_{\lambda\mu}$ and **(ii)** $\mathrm{Ind}\,(\varepsilon C_{\lambda\mu})\!\uparrow^{B_n}_{C_{\lambda\mu}} \cong K[B_n]b_{\lambda\mu}$.   ◇

Finally, we conclude with the following result which implies that the FOUR realisations of the irreducible representations for B_n, namely, **(I)** $U_{(\lambda,\mu)}$, **(II)** $V_{(\lambda,\mu)}$, **(III)** $W_{(\lambda,\mu)}$ and **(IV)** $[\lambda,\mu]$ are all equivalent for every $(\lambda,\mu) \models n$.

7.9.5 Theorem: *For all* $(\lambda,\mu) \models n$, *we have* $[\lambda,\mu] \cong U_{(\lambda,\mu)}$.

Proof: With the notation as in (7.3.2) and (7.3.3) above, we have $U_{(\lambda,\mu)} = \left(\chi_{\ell m} \otimes (V_\lambda \otimes V_\mu)\right)\!\uparrow^{B_n}_{I_{\ell m}}$. We have only to show that the irreducible representation $U_{(\lambda,\mu)}$ occurs in both $\mathrm{Ind}\,(\iota R_{\lambda\mu})\uparrow^{B_n}_{R_{\lambda\mu}}$ and $\mathrm{Ind}\,(\varepsilon C_{\lambda\mu})\uparrow^{B_n}_{C_{\lambda\mu}}$. This is a routine verification using the reciprocity and the subgroup theorems.   ◊

7.9.6 Corollary: *Using (7.3.6) above, we have*
(i) $[\mu,\lambda] \cong [\lambda,\mu] \otimes_K (\xi B_n)$ *and consequently,*
(ii) $[\lambda,\mu]{\downarrow}^{B_n}_{D_n} \cong [\mu,\lambda]{\downarrow}^{B_n}_{D_n}$ *(where $D_n = \mathrm{Ker}\,(\xi)$).*   ■

7.10   Exercises

1. Show that $B_n = N \bullet S_n$ is the semi–direct product (Ex.(4.8.11) above) of S_n by the abelian group $N = C_2^n$ (for the natural action of S_n on N by permuting coordinates) where $C_2 = \{\pm 1\}$ is the cyclic group of order 2.

2. Let η be the character of B_n, as defined in (7.1.6) above. Show that $\mathrm{Ker}(\eta) = N \bullet A_n \subseteq N \bullet S_n$. Are the other two subgroups of index 2 in B_n, namely, $D_n = \mathrm{Ker}(\xi)$ and $\mathrm{Ker}(\varepsilon)$, also semi–direct products of suitable groups?

   The group $\mathrm{Ker}(\eta)$ is called a *generalised alternating group* and is denoted by $A_n^\star$. See [41] and [51] for a study of the representations of the so called *"generalised symmetric groups"*.

3. List the conjugacy classes and construct the character tables for B_4 and B_5.

4. Using the method of little groups, describe $\mathrm{Irr}_K(A_n^\star)$.

5. Using *only* the results of § 7.3 above, show that
   $$U_{(\mu',\lambda')} \cong U_{(\lambda,\mu)} \otimes_K \varepsilon B_n \text{ for all } (\lambda,\mu) \models n.$$

6. Using the formula (5.2.6) above, find the number of elements in the conjugacy class in B_n of cycle type $(\lambda,\mu) \models n$.   ■

Chapter 8

Representations of the Hyperoctahedral Group D_n

In this chapter, we shall define the hyperoctahedral group of type D_n which is a subgroup of B_n of index 2 (8.1.0) and follow the outlay of Chapter 6 replacing A_n by D_n and S_n by B_n to determine all (ordinary) irreducible representations of D_n. The considerations for the case of (B_n, D_n) are similar to (S_n, A_n) and so we need only to set the notation and state the results. (Cf. [37], [48], etc.)

8.1 The Hyperoctahedral Group D_n

Recall the definition of the hyperoctahedral group of type B_n (7.1.0) which is a subgroup of S_{2n} given by

$$B_n = \{\ \theta \in S_{2n} \ | \ \theta(i) + \theta(-i) = 0, \ \forall \ i \ , 1 \leq i \leq n\}$$

where S_{2n} is the permutation group on the $2n$ symbols $\pm 1, \cdots, \pm n$.

8.1.0 The group D_n: For an integer $n \geq 4$, the *hyperoctahedral group of type D_n* or simply the *hyperoctahedral group D_n* (of rank n) is defined to be the subgroup of B_n given by

$$D_n = \{\theta \in B_n \ | \ \text{Card}\,\{i \ | \ \theta(i) \ < \ 0, 1 \leq i \leq n\} \text{ is even }\},$$

or equivalently, $D_n = \{\ \theta \in B_n \ | \ \theta(1) \cdots \theta(n) \ > \ 0\}$. It is a simple matter to verify that D_n is a subgroup of B_n.

8.1.1   The order of D_n: Recall that an element θ of D_n is of the form (7.1.1),

$$\theta = \begin{pmatrix} 1 & \cdots & n & -n & \cdots & -1 \\ \epsilon_1\sigma(1) & \cdots & \epsilon_n\sigma(n) & -\epsilon_n\sigma(n) & \cdots & -\epsilon_1\sigma(1) \end{pmatrix}$$

for unique $(\epsilon_1, \cdots, \epsilon_n) \in C_2^n$ and $\sigma \in S_n$ such that $\epsilon_1 \cdots \epsilon_n = 1$ where $C_2 = \{\pm 1\}$ is the cyclic group of order 2 and S_n is the symmetric group on $1, \cdots, n$. Hence D_n is of order $2^{n-1}n!$. Yet another way to see this is as follows.

Let $\theta \in B_n$ have its cycle decomposition ((7.1.2) and(7.2.1)),

$$\theta = \theta_1\bar{\theta}_1\theta_2\bar{\theta}_2 \cdots \theta_r\bar{\theta}_r\vartheta_1\vartheta_2 \cdots \vartheta_s,$$

where $\theta_j\bar{\theta}_j$ is a positive cycle of length λ_j $(1 \le j \le r)$ and ϑ_k is a negative cycle of length μ_k $(1 \le k \le s)$. Then it is easy to see that $\theta \in D_n$ if and only if s is *even*. In particular, we have $D_n = \mathrm{Ker}\,\xi$ where ξ is the character of the group B_n, taking the values ± 1, as defined in (7.1.6) above. Hence D_n is a subgroup of index 2 in B_n.■

8.2   Conjugacy Classes of D_n

Let $\theta \in D_n$ be of cycle type $(\lambda, \mu) \models n$ (as an element of B_n) and let $C_{(\lambda,\mu)}^{B_n}(\theta)$ be its conjugacy class in B_n. Recall (§ 4.6) that $C_{(\lambda,\mu)}^{B_n}(\theta) \subseteq D_n$ and it *remains one* or *splits* into a union $C_{(\lambda,\mu)}^+ \cup C_{(\lambda,\mu)}^-$ of *two* conjugacy classes of D_n.

8.2.1   Theorem: *With notation as above, we have the following.*
1. $C_{(\lambda,\mu)}^{B_n}(\theta)$ *remains a conjugacy class of D_n if and only if* <u>either</u> $\mu \ne 0$ <u>or else</u> *one of the parts of λ is odd. Or equivalently,*
2. $C_{(\lambda,\mu)}^{B_n}(\theta)$ *splits into a union $C_{(\lambda,\mu)}^+ \cup C_{(\lambda,\mu)}^-$ of two conjugacy classes of D_n if and only if $\mu = 0$ <u>and</u> all the parts of λ are even.*
In particular, if n is odd, the conjugacy class of any element θ in the subgroup D_n is the same as the conjugacy class of θ in B_n.

Proof: The proof is divided into four steps.

Step 1: For $\theta \in D_n$, let $C_{D_n}(\theta)$ and $C_{B_n}(\theta)$ be the centralisers of θ in D_n and B_n respectively. By (4.6.1) above, we have only to show that

$$C_{D_n}(\theta) \neq C_{B_n}(\theta) \iff \begin{cases} \mu \neq 0 \text{ or} \\ \mu = 0 \text{ and } (-1)^{\lambda_i} = -1 \text{ for some } i. \end{cases}$$

Or equivalently,

$$C_{B_n}(\theta) \not\subseteq D_n \iff \begin{cases} \mu \neq 0 \text{ or} \\ \mu = 0 \text{ and } (-1)^{\lambda_i} = -1 \text{ for some } i. \end{cases}$$

Let $\theta \in B_n$ have its cycle decomposition (7.2.1),

$$\theta = \theta_1 \overline{\theta}_1 \theta_2 \overline{\theta}_2 \cdots \theta_r \overline{\theta}_r \vartheta_1 \vartheta_2 \cdots \vartheta_s,$$

where $\theta_j \overline{\theta}_j$ is a positive cycle of length λ_j, $1 \leq j \leq r$ and ϑ_k is a negative cycle of length μ_k, $1 \leq k \leq s$. Since $\theta \in D_n$, we have s is even. Our goal is to show that

$$C_{B_n}(\theta) \not\subseteq D_n \iff \begin{cases} s \geq 1 \text{ or} \\ s = 0 \text{ and } (-1)^{\lambda_i} = -1 \text{ for some } i. \end{cases}$$

Step 2: If $s \geq 1$, then $\vartheta_1 \in C_{B_n}(\theta) - D_n (\Rightarrow C_{B_n}(\theta) \not\subseteq D_n)$.

This is obvious since no negative cycle can be in D_n and hence in particular $\vartheta_1 \notin D_n$ and further, any element commutes with its cycle factors and so $\theta \vartheta_1 = \vartheta_1 \theta$, as required.

Step 3: Suppose $\mu = 0$ and λ_i is *odd* for some i. Then $\sigma \in C_{B_n}(\theta) - D_n$ where $\sigma = (a_1, -a_1) \cdots (a_{\lambda_i}, -a_{\lambda_i})$ and $\theta_i = (a_1, \cdots, a_{\lambda_i})$. Since σ is a product of odd number of negative cycles, it cannot be in D_n. On the other hand, σ commutes with θ since $\sigma \theta_i \overline{\theta}_i(a_j) = \sigma \theta_i(a_j) = \sigma(a_{j+1}) = -a_{j+1} = \theta_i(-a_{j+1}) = \theta_i \overline{\theta}_i \sigma(a_j)$.

Step 4: $C_{B_n}(\theta) \subseteq D_n$ if $\mu = 0$ and λ_i's are *even* for all i.

If $\sigma \in B_n$ commutes with $\theta = \theta_1 \overline{\theta}_1 \cdots \theta_r \overline{\theta}_r$, with all the θ_i's of even length, then we have to show that $\sigma \in D_n$.

By the uniqueness of the cycle decomposition, σ permutes (by conjugation) the cycle factors of θ of the same length . There are two possibilities: **(i)** either σ conjugates a θ_j to another θ_k, or

(ii) σ conjugates a θ_j to a $\overline{\theta}_k$.

Suppose $\lambda_{i-1} > \lambda_i = \lambda_{i+1} = \cdots = \lambda_{i+\ell} > \lambda_{i+\ell+1}$. We may assume that **(i)** σ conjugates θ_i's cyclically, i.e., $\theta_i \mapsto \theta_{i+1} \mapsto \cdots \mapsto \theta_{i+\ell} \mapsto \theta_i$, or **(ii)** σ conjugates a θ_i to $\overline{\theta}_i$. Let us assume that

$$
\begin{aligned}
\theta_i &= (a_{i1}, \quad \cdots, \quad a_{i\lambda_i}) \\
\theta_{i+1} &= (a_{(i+1)1}, \quad \cdots, \quad a_{(i+1)\lambda_i}) \\
\vdots \quad &\qquad \vdots \qquad \cdots, \qquad \vdots \\
\theta_{i+\ell} &= (a_{(i+\ell)1}, \quad \cdots, \quad a_{(i+\ell)\lambda_i})
\end{aligned}
$$

Now it is easy to see that in the case **(i)** we have

$$
\begin{aligned}
\sigma = \ &(a_{i1}, a_{(i+1)1}, \cdots, a_{(i+\ell)1})(-a_{i1}, -a_{(i+1)1}, \cdots, -a_{(i+\ell)1}) \cdots \\
&\cdots (a_{i\lambda_i}, a_{(i+1)\lambda_i}, \cdots, a_{(i+\ell)\lambda_i})(-a_{i\lambda_i}, -a_{(i+1)\lambda_i}, \cdots, -a_{(i+\ell)\lambda_i})
\end{aligned}
$$

which is a product of only positive cycles and hence $\sigma \in D_n$.

In the case **(ii)**, we see that $\sigma(a_{jk}) = -a_{jk}$ for all j and k which implies that in the cycle decomposition of σ, the even number of negative transpositions $(a_{jk}, -a_{jk})$ $(k = 1, \cdots, \lambda_j)$ occur since λ_j is even. Thus σ is a product of certain positive cycles and an even number of negative cycles and hence $\sigma \in D_n$, as required. ∎

8.3   Irreducible Representations of D_n

Given a complete set of irreducible representations of B_n, (say for example, as in §7.3 above), $\mathrm{Irr}_K(B_n) = \{U_{(\lambda,\mu)} \mid (\lambda, \mu) \models n\}$, we get all the irreducible representations of D_n (using (4.6.5) above), once we know how to classify the $(\lambda, \mu) \models n$ into the *two cases* that $U_{(\lambda,\mu)} \downarrow_{D_n}^{B_n}$ *remains irreducible* or *not* as a D_n-module. We are in a position to achieve this, by essentially repeating what we have done (in §6.3 above) for the case of A_n. The method and the proofs are identical with one small difference. We shall merely sketch an outline setting the parallel notation.

The difference is that the role played by the relation (6.2.3), $[\lambda] = [\lambda'] \otimes (\varepsilon S_n)$ (in the case of A_n), is to be replaced by (7.3.6) above, i.e., $U_{(\mu,\lambda)} = U_{(\lambda,\mu)} \otimes (\xi B_n)$.

Notation: Given an irreducible representation (V, ρ) of B_n, we write
(i) $V^0 = V\!\downarrow_{D_n}^{B_n}$ if the latter is irreducible for D_n, or
(ii) $V^+ \oplus V^- = V\!\downarrow_{D_n}^{B_n}$ if the latter is reducible (in which case it decomposes into two inequivalent irreducible representations).

Remark: Since $U_{(\lambda,\mu)}\!\downarrow_{D_n}^{B_n} = U_{(\mu,\lambda)}\!\downarrow_{D_n}^{B_n}$ (7.3.6), whenever the objects $U_{(\lambda,\mu)}^0$, $U_{(\lambda,\mu)}^\pm$ are defined for (λ, μ) so are the similar ones for (μ, λ) and vice–versa. In fact, we have

$$U_{(\lambda,\mu)}^0 = U_{(\mu,\lambda)}^0 \text{ and } U_{(\lambda,\mu)}^\pm = U_{(\mu,\lambda)}^\pm, \ \forall \ (\lambda, \mu) \models n.$$

8.3.1 Theorem: *For all $(\lambda, \mu) \models n$, we have the following.*
1. $U_{(\lambda,\mu)}^0 = U_{(\lambda,\mu)}\!\downarrow_{D_n}^{B_n}$ *is irreducible for D_n $\iff \lambda \neq \mu$.*
2. $U_{(\lambda,\mu)}\!\downarrow_{D_n}^{B_n} = U_{(\lambda,\mu)}^+ \oplus U_{(\lambda,\mu)}^- \iff \lambda = \mu$.
3. *The three kinds of irreducible representations $U_{(\lambda,\mu)}^0$ with $\lambda \neq \mu$ and $U_{(\lambda,\lambda)}^\pm$ are mutually inequivalent.*
4. *Every irreducible representation of D_n is equivalent to one of the three kinds above, i.e., the family*

$$\mathrm{Irr}_K(D_n) = \{U_{(\lambda,\mu)}^0 \mid \lambda \neq \mu\} \bigcup \{U_{(\lambda,\lambda)}^\pm \mid \lambda \vdash \ell, 2\ell = n\}$$

is a complete set of irreducible representations of D_n.
In particular, if n is <u>odd</u>, the family $\{U_{(\lambda,\mu)}^0\}$, parametrised by the <u>unordered pairs</u> (λ, μ) (of complementary partitions of n) is a complete set of irreducible representations of D_n.

Proof: The proof is identical with that of (6.3.1) above. It suffices therefore to sketch an outline.

Step 1: Every $(\lambda, \mu) \models n$ gives rise to one or two of the irreducible B_n–modules (W_i, ϑ_i), $i = 1, 2$ and 3, as defined below.

Let (W, ϑ) be an irreducible component of $U_{(\lambda,\mu)}\!\downarrow_{D_n}^{B_n} = U_{(\mu,\lambda)}\!\downarrow_{D_n}^{B_n}$ and I_W be its inertia group (4.3.5). We have $I_W \supseteq D_n$ and further
$I_W = B_n \iff W$ is self-conjugate, i.e., $\vartheta \cong \vartheta^{(1,-1)}$ where $\vartheta^{(1,-1)}((1,-1)\sigma(1,-1)) = \vartheta(\sigma), \ \forall \ \sigma \in D_n$.

Case I: Suppose W is self–conjugate.

Extend (W, ϑ) to a representation $(W, \tilde{\vartheta})$ of B_n by defining $\tilde{\vartheta}(\sigma) = \tilde{\vartheta}(\sigma(1, -1)) = \vartheta(\sigma)$ for all $\sigma \in D_n$. This gives rise to two irreducible inequivalent representations of B_n, namely,

$$(W_1, \vartheta_1) = (W, \tilde{\vartheta}) \otimes (\iota B_n) = (W, \tilde{\vartheta}) \quad \text{and} \quad (W_2, \vartheta_2) = (W, \tilde{\vartheta}) \otimes (\xi B_n).$$

We have $(\star)$: $\mathrm{Ind}\,(W){\uparrow}_{D_n}^{B_n} \cong W_1 \oplus W_2$ and $U_{(\lambda,\mu)} \cong W_1$ or W_2.

Case II: Suppose W is <u>not</u> self–conjugate.

We have $U_{(\lambda,\mu)}{\downarrow}_{D_n}^{B_n} = (W, \vartheta) \oplus (W, \vartheta^{(1,-1)})$ and $I_W = D_n$ and hence (W, ϑ) or its conjugate $(W, \vartheta^{(1,-1)})$ gives rise to the irreducible representation of B_n, $(\star\star)$: $(W_3, \vartheta_3) = \mathrm{Ind}\,(\vartheta){\uparrow}_{D_n}^{B_n} \cong U_{(\lambda,\mu)}$. By $(\star)$ and $(\star\star)$, we have shown the following.

Step 2: Every irreducible representation $U_{(\lambda,\mu)}$ of B_n is equivalent to (W_j, ϑ_j) for some $j = 1,\ 2$ or 3.

Finally, we have only to prove the following.

Step 3: (i) $U_{(\lambda,\mu)} \cong W_1$ or $W_2 \iff (W, \vartheta)$ is self–conjugate $\iff$ $\lambda \neq \mu$, or equivalently,
(ii) $U_{(\lambda,\mu)} \cong W_3 \iff (W, \vartheta)$ is <u>not</u> self–conjugate $\iff \lambda = \mu$.

This is a consequence of the symmetry of the situation that whatever is done above remains the same if (λ, μ) is replaced by (μ, λ), etc. ∎

8.4   Exercises

1. Count the number $h(D_n)$ of conjugacy classes of D_n in terms of $h(B_n)$.

2. Give a bijection between the set of conjugacy classes of D_n with a complete set of mutually inequivalent irreducible representations of D_n (8.3.3). Is the task any simpler in the case when n is *odd?*

3. List the conjugacy classes and construct the character table of D_4. Do the same for D_5.

4. Compare and contrast the information gathered for the groups B_n and D_n, similar to what was done for S_n and A_n (in §6.4 above). ∎

Bibliography

[1] AL-Aamily, A.O. Morris and M.H. Peel, The Representations of the Weyl Groups of Type B_n, J. Algebra, **68**(1981), 298–305.

[2] K. Akin, Representations of GL(n) and Schur Algebras, *Commutative Algebra and Combinatorics*, Adv. Studies in Pure Math. **11**(1987), 1–8.

[3] G.E. Andrews, *The Theory of Partitions*, Encyclopaedia of Math. and its Appl. Vol.**2**, Addison–Wesley Publishing Company, London (1976).

[4] F.Bergeron and N. Bergeron, A Decomposition of the Descent Algebra of the Hyperoctahedral Groups–I, J. of Algebra, **148**(1992), 86–97.

[5] N. Bergeron, An Hyperoctahedral Analogue of the Free Lie Algebra, J. Combin. Theory, **58**(1991), 256–278.

[6] N. Bergeron, A Decomposition of the Descent Algebra of the Hyperoctahedral Groups–II, J. of Algebra, **148**(1992), 98–122.

[7] R.W. Carter, Conjugacy Classes in the Weyl Group, *Seminar on Algebraic Groups and Related Finite Groups*, Lecture Notes in Math. No.**131**, Springer (1970).

[8] R.W. Carter and G. Lusztig, On the Modular Representations of the General Linear and Symmetric Groups, Math. Zeit. **136**(1974), 193–242.

[9] C.W. Curtis and T.V. Fossum, On Centraliser Rings and Representations of Finite Groups, Math. Zeit. **107**(1968), 402–406.

[10] C.W. Curtis and I. Reiner, *Representation Theory of Finite Groups and Associative Algebras*, Pure and Applied Math. Texts and Monographs–Vol.XI, John Wiley and Sons, New York (1962).

[11] C.W. Curtis and I. Reiner, *Methods of Representation Theory with Applications to Finite Groups and Orders*, Vol.I, John Wiley and Sons, New York (1981).

[12] J. Désarménien, J.P.S. Kung and G.-C. Rota, Invariant Theory, Young Bitableaux and Combinatorics, Adv. in Math. **27**(1978), 63–92.

[13] S. Donkin, On Schur Algebras and Related Algebras–I and II, J. Algebra **104**(1986), 310–328 and **111**(1987), 354–364.

[14] H.K. Farahat and M.H. Peel, On the Representation Theory of Symmetric Groups, J. Algebra, **67**(1980), 280–304.

[15] J.S. Frame, Orthogonal Group Matrices of Hyperoctahedral Groups, Nagoya Math. Journal, **27**(1966), 585–590.

[16] A. Garsia and C. Reutenauer, A Decomposition of Solomon's Descent Algebra, Adv. in Math. **77**(1989), 189–262.

[17] A. Garsia and McLarnan, Relations between Young's Natural and the Kazhdan–Lusztig Representations of S_n, Adv. in Math. **69**(1988), 32–92.

[18] L. Geissinger and D. Kinch, Representations of the Hyperoctahedral Groups, J. Algebra, **53**(1978), 1–20.

[19] J.A. Green, *Polynomial Representations of* GL_n, Lecture Notes in Math. No.**830**, Springer (1980).

[20] E.A. Gutkin, Representations of the Weyl Group in the Zero Weight Space, Uspekhi Mat. Nauk (no. **5(173)**), **28**(1973), 237–238 (in Russian), English Summary in **MR. 53**, # 13423.

[21] R. Howe, The Classical Groups and Invariants of Binary Forms, Proc. of Symposia in Pure Math. Vol.**48**: *The Mathematical Heritage of Hermann Weyl*, Amer. Math. Soc. Providence, Rhode Island (1987).

[22] J.E. Humphreys, *Intoduction to Lie Algebras and Representation Theory*, Graduate texts in Math. Vol.**9**, Springer (1972).

[23] N. Jacobson, *Basic Algebra* I and II, Hindustan Publishing Corporation (India), New Delhi (1984).

[24] G.D. James, *The Representation Theory of the Symmetric Group*, Lecture Notes in Math. No.**682**, Springer (1978).

[25] G. James and A. Kerber, *The Representation Theory of the Symmetric Group*, Encyclopaedia of Math. and its Appl. Vol.16, Addison–Wesley Publishing Company (1981).

[26] G. Karpilovsky, *Induced Modules over Group Algebras*, North–Holland Math. Studies **161**, Amsterdam (1990).

[27] R. Keown, *An Introduction to Group Representation Theory*, Academic Press, New York (1975).

[28] A. Kerber, *Representations of Permutation Groups–1*, Lecture Notes in Math. No.240, Springer (1971).

[29] D.E. Knuth, Permutations, Matrices and Generalised Young Tableaux, Pacific J. Math. **34**(1970), 709–727.

[30] K. Koike and I. Terada, Young–Diagramatic methods for the Representation Theory of the Classical Groups of Type B_n, C_n and D_n, J. Algebra, **107**(1987), 466–511.

[31] S. Lang, *Algebra*, (2nd ed.), Addison Wesley, London (1984).

[32] W. Ledermann, *Introduction to Group Characters*, Cambridge Univ. Press, Cambridge (1977).

[33] I.G. Macdonald, Some Irreducible Representations of Weyl Groups, Bull. Lond. Math. Soc. **4**(1972), 148–150.

[34] I.G. Macdonald, On the Degrees of the Irreducible Representations of Finite Coxeter Groups, J. Lond. Math. Soc. (2)**6**(1973), 298–300.

[35] I.G. Macdonald, *Symmetric Functions and Hall Polynomials*, Clarendon Press, Oxford (1979).

[36] S.J. Mayer, On the Irreducible Characters of the Symmetric Group, Adv. in Math. **14**(1974), 127–132.

[37] S.J. Mayer, On the Characters of Weyl Groups of Type D, Math. Proc. Camb. Phil. Soc. **77**(1975), 259–264.

[38] W. Miller Jr., A Branching Law for the Symmetric Groups, Pacific J. Math. **36**(1966), 341–346.

[39] W. Miller Jr., *Symmetry Groups and their Applications*, Academic Press, New York (1972).

[40] C. Musili, *Introduction to Rings and Modules*, Narosa Publishing House, New Delhi (1991).

[41] M. Osima, On the Representations of Generalised Symmetric Group, Math. J. Okayama Univ. 4(1954), 39–56.

[42] D.S. Passman, *The Algebraic Structure of Group Rings*, Pure and Applied Math. Texts, Monographs and Tracts, John Wiley and Sons, New York (1977).

[43] M.H. Peel, Hook Representations of the Symmetric Groups, Glasgow Math. J. 12(1971), 136–149.

[44] M.H. Peel, Specht Modules and Symmetric Groups, J. Algebra, 36(1975), 88–97.

[45] C. Reutenauer, Theorem of Poincaré Birkhoff Witt and Symmetric Group Representations of Degrees equal to Stirling Numbers, in Lecture Notes in Math. No. 1234, Spinger (1986), 267–293.

[46] J.-P. Serre, *Linear Representations of Finite Groups*, Graduate Texts in Math. Vol.42, Springer (1977).

[47] T.A. Springer, A Construction of Representations of Weyl Groups, Invent. Math. 44(1978), 279–293.

[48] G. Srinagesh, Representations of Classical Weyl Groups, M.Phil. Dissertation, Univ. of Hyderabad, Hyderabad (1987–88).

[49] M. Sundari, Representations of GL_n, M.Phil. Dissertation, Univ. of Hyderabad, Hyderabad (1988–89).

[50] T I F R, *Semi-simple Rings*, Summer School Notes (Unpublished), TIFR, Bombay (1969).

[51] A. Young, On Quantitative Substitutional Analysis (Paper V), Proc. Lond. Math. Soc. 31(1930), 273–288.

Index

Texts and Readings in Mathematics